SPACE SCIENCE, EXPLORATION AND POLICIES

SPACE ROBOTICS SUPPORTING EXPLORATION MISSIONS

VISION, FORCE CONTROL AND COORDINATION STRATEGIES

SPACE SCIENCE, EXPLORATION AND POLICIES

Additional books in this series can be found on Nova's website under the Series tab.

Additional e-books in this series can be found on Nova's website under the e-book tab.

SPACE SCIENCE, EXPLORATION AND POLICIES

SPACE ROBOTICS SUPPORTING EXPLORATION MISSIONS

VISION, FORCE CONTROL AND COORDINATION STRATEGIES

ENRICA ZEREIK

nova publishers
New York

For permission to use material from this book please contact us:
Telephone 631-231-7269; Fax 631-231-8175
Web Site: http://www.novapublishers.com

NOTICE TO THE READER

Library of Congress Cataloging-in-Publication Data

Space robotics supporting exploration missions : vision, force control and coordination strategies / editor, Enrica Zereik.
p. cm.
Includes bibliographical references and index.
ISBN 978-1-62417-894-8 (hardcover)
1. Space robotics. 2. Robot vision. I. Zereik, Enrica.
TL1097.S63 2013
629.46--dc23
2012051462

Published by Nova Science Publishers, Inc. † New York

Contents

Acknowledgments

First of all, I would like to really thank Professor Giuseppe Casalino for giving me the great opportunity to accomplish my Ph.D. studies and for teaching me a lot of things, from both a scientific and a human point of view.

Sincere thanks go also to Alessio Turetta and Andrea Sorbara, who very often supported me with their knowledge and patience, and also thanks to all lab members and friends, for both help and laughs in the years spent at Graal.

Many thanks to ESA staff, particularly to Mr. Frederic Didot, for the very interesting project I had the honour to work on.

And, last but not least, I would like to acknowledge also the entire Alenia crew, for highly supporting me with their advice and friendship in long days far from home.

List of Symbols

BP	Backward Phase
CCD	Charge-Coupled Device
CNES	Centre Nationale d'Etudes Spatiales
CSA	Canadian Space Agency
d.o.f.	degree of freedom
DEM	Digital Elevation Model
DP	Dynamic Programming
DSS	Dry Segment Subsystem
EAC	European Astronaut Centre
EDL	Entry Descent and Landing
EFD	Eurobot Flight Demonstrator
EFM	Eurobot Flight Model
EGNC	Eurobot Guidance and Navigation Control
EGP	Eurobot Ground Prototype
ESA	European Space Agency
ETPSE	EGP Test Playground Support Equipment
EVA	Extra Vehicular Activity

EWM	Eurobot Wet Model
ExoMars	Exobiology on Mars
FAST	Features from Accelerated Segment Test
FP	Forward Phase
FPGA	Field Programmable Gate Array
fps	frames per second
GEO	Geostationary Earth Orbit
GigE	Gigabit Ethernet
HSV	Hue Saturation Value
ISS	International Space Station
IVA	Intra Vehicular Activity
JAXA	Japan Aerospace Exploration Agency
JEM	Japanese Experiment Module
LEO	Low Earth Orbit
lidar	Light Detection and Ranging
LQ	Linear Quadratic
MEO	Medium Earth Orbit
MER	Mars Exploration Rovers
MLE	Maximum Likelihood Estimation
MLI	Multi-Layer Insulation
MM	Manipulability Measure
MMI	Man-Machine Interface
MSL	Mars Science Laboratory

NASA	National Aeronautics and Space Administration
NBF	Neutral Buoyancy Facility
ORU	Orbital Replaceable Unit
PRR	Preliminary Requirements Review
PTU	Pan/Tilt Unit
RANSAC	RANdom SAmple Consensus
RCS	Robotic Control System
RGNC	Rover Guidance and Navigation Control
RKA	Russian Federal Space Agency
RMS	Remote Manipulator System
SIFT	Scale-Invariant Feature Transform
sol	solar day on Mars
SPDM	Special Purpose Dexterous Manipulator
SURF	Speeded Up Robust Features
SUSAN	Smallest Univalue Segment Assimilating Nucleus
SVD	Singular Value Decomposition
VPU	Vision Processing Unit
VRML	Virtual Reality Modeling Language
WET	Weightless Environmental Test
WSS	Wet Segment Subsystem

List of Figures

List of Tables

Preface

In recent years interest in space activities has grown more and more and there are future space exploration plans that include human missions to the Moon and Mars; the return to the Moon will rely on a large robotic presence to support exploration and, later, to assist the building of the lunar base.

These plans will thus require the availability of intelligent, interactive and cooperative robotic systems since, on space missions, astronaut time is valuable and very scarce. Any help that autonomous robotic platforms can provide humans with (for example in order to accomplish routine, repetitive or dangerous tasks) will free the astronauts to focus on research that requires their expertise. So there is the need of developing skills, capabilities and operational experience for working in space and on the surface of the Moon (and Mars later) with machines and people, in preparation of further destinations.

In order to gain a competitive role in this exploration scenario, one of the ESA major interests is that of strengthen the European robotic knowledge and competence for space applications. More in details, ESA is interested in the creation and maintenance of an industrial technology base for the automation and remote control of space based operations. As such, the domain of required competence includes the specification and control of space robotics systems, from manipulators to autonomous vehicles, and the support to automation and robotics in manned and unmanned missions. This competence acquisition can be accomplished by developing in parallel streams the enabling technologies, the requirements for space robotics and robotic system demonstrators. Space Robotics is a highly multidisciplinary subject: it heavily depends on space issues such as structures, mechanisms, avionics, on board software, ground operations and communication. However, packing tightly together all these disciplines in a single system that can move, interact physically with the surrounding environment and with humans, while showing a certain degree of autonomy, introduces a high level of complexity.

In order to specify, design and ultimately validate space automation and robotic systems that fulfill astronaut needs, under defined environmental and operational constraints, specific expertise are needed for:

- definition of utilization concepts for mission success;
- functional, kinematic and dynamic analysis;
- physical or simulated modeling of robots;
- testing of robotic platforms;
- development of effective interfaces for robot-human interaction (e.g. haptic devices);
- robot control, both autonomy and teleoperation;
- analysis of vision systems;
- communication management.

In particular, the purpose of the demonstrators is to integrate the technologies made available from other studies together with existing commercial technologies in a working system to be utilized as a testbed for early concept evaluation. A physical "living" prototype, built and tested according to a scenario as much as possible representative, allows the engineers to identify and solve some of the arising critical situations at an early stage and most of all gives the possibility to the users, like astronauts, to participate to the design process by attending tests and giving feedback, thus steering the design choices from the start of the project.

In order to give to these robotic systems the high level of autonomy they need to be useful in space exploration missions, one of the main topics is to provide the robot with the capability to interact with the environment, thus being able to perform both cooperative tasks (helping the astronauts) and autonomous operations (that is tasks not requiring human participation). This Thesis fits in this global scenario as it is intended to present some results obtained within this research on cooperative and autonomous robotic systems for space exploration missions.

Chapter 1

Introduction

1.1. Motivations and Previous Missions

Man has always shown great interest in space exploration, both for reasons linked to the knowledge and the comprehension of the universe origin, and, above all in the last years of his technological evolution, in order to evaluate the capabilities of different other planets to sustain a permanent human-robotic colonized presence. In such a way to achieve these purposes, innovative and more and more "smart" technologies have been used, in order to achieve the objective of performing the required tasks in an autonomous and independent way, without the involvement of a human being.

Robots and automated systems assisting human astronauts have been used in space missions in which automation and robotics played a substantial role in installing infrastructures that were essential for space exploration: both Earth-orbiting satellites and space stations occupied for extended periods by astronauts themselves.

The space environment poses robotics in front of unique challenges in order to develop suitable control algorithms, considering also the arising interest in the opportunities for human-robot and robot-robot cooperation; particularly, some of the priority areas, subject to scientific and technological research, include robotic locomotion and manipulation under gravity conditions that differ from those on the Earth, the realization of effective control interfaces (including teleoperated modes), the analysis and compensation for the effects that glare and glint have on machine vision...

The use of robotic technology cannot be separated from some aspects, essential for space mission success, such as:

1. *safety*: it is possible to relieve the astronauts of some operations that can be extremely dangerous, for example because of the hostile environment they have to work into;

2. *performance*: tasks too hard or even impossible for a human being (for example for the high precision and repeatability required in the execution or for the long duration of the operation itself) can be executed by a proper robotic system;

3. *cost*: astronaut presence in space requires very expensive life support infrastructures and, moreover, humans must return to Earth, while robots just need power during their working period and then they can be simply disposed of after they have attained their goals.

Space robots are, in many ways, very different from terrestrial robots and their design is driven by particular constraints and requirements that can be grouped into two different classes: space environmental constraints and space system constraints, linked to the specific mission and to its purpose.

For instance, some of the guidelines arising from the particular environment in which these systems have to work into, result in requirements about a careful material selection and a special design, supplying to some lacks in the usable technology: in fact, certain sensing principles, as the ultrasonic, are not applicable in the control of the interaction between the system and the environment. Besides, from the extreme environmental conditions, such as the exposure to high levels of radiation, some problems arise: both the limited material lifetime and the outdated computer performance, since not all kind of hardware can be used. In fact the only technology available for space use is the *space qualified* one: exploration missions can only rely on computers with worse performances than the state-of-the-art ones. Furthermore, vision and image processing, too, already damaged by the few available computational resources, are made more difficult by the extreme lighting and contrast conditions that characterize space. Finally, as they must work in an extremely remote environment, these robotic systems must be endowed with an adequate level of autonomy both for what concerns their behavior and the operations they must execute; also their being, as much as possible, maintenance-free systems turns out to be an important feature. Another very important aspect is the testing phase before launch. Since communications with Earth come to be very limited and expensive, they are often not enough in order to guide the overall execution of a task. Therefore all the possible situations the system may get involved in must be carefully experienced and, in particular, the robot behavior in such situations must be thoroughly checked. Sometimes, anyhow, environmental constraints may also have positive design impacts: this is the case of robotic systems physically impossible on Earth (e.g. small robots with the ability to manipulate enormous payloads, thanks to reduced gravity).

Further system constraints, essential in order to obtain a correct robotic behavior and to achieve mission success, combine to bring about the final design of the system: first of all, each robot to be employed in space missions has to fulfill requirements of high reliability and safety, in order to ensure both a correct execution of its own tasks and the accomplishment of its own goals, and, above all, in order to obtain a high level of confidence in the safety of the interaction with astronauts, preventing the robot itself to be a further source of danger for them. Moreover, as already suggested, as maintenance during the mission is quite difficult if not impossible, a maintenance-free design is essential, together with an upgrade ability constraint (e.g. re-programmability), in order to endow the system with the longest possible lifetime and to decrease the number of required maintenance actions (to which, however, the so-called ORU - "orbital replaceable units" - are devoted).

A remarkable critical point in the design of a robotic system for space applications is the availability of a technology that is often obsolete with respect to the goals that space exploration missions propose to reach: only space qualified hardware can be employed in space applications, in order to comply with the requirements of safety and high reliability of the system itself; moreover, the very limited and expensive on board power contributes to limit both the available computing resources and the on board control electronics. In fact this is one of the major problems encountered by space robotics: not all knowledge and hardware employed by terrestrial robotics can be exploited in space missions. This is limiting because the most recent technologies cannot be considered, above all because they are not very reliable and robust: autonomous systems with complex behavior are, in fact, hard to characterize in order to guarantee minimal performance criteria under all reasonable circumstances.

Furthermore, because of the very poor and expensive possibility to communicate with Earth during missions, it is essential to add a further requirement of an adequate degree of autonomy for the system but also to realize very sophisticated ground operator interfaces that could effectively control robots whenever communication is possible and that could cope with problems such as signal transmission delays.

There are mainly two different fields in space robotics, concerning different scenarios in which robotic systems could be employed: *"Orbital Robotics"* and *"Planetary Robotics"*. With regard to the first one, although the use of robotic systems is technically possible in the main orbital domains, such as the *Geostationary Earth Orbit* (*GEO*), base of the commercial/military communications and Earth observation satellites, or the *Medium Earth Orbit* (*MEO*), base of the Global Positioning System satellites, robots are solely used in the *Low Earth Orbit* (*LEO*). This last, from 300 to 700 km of altitude, is the domain of manned space missions: examples of such missions are the *International Space Station* (*ISS*) and the past *US Skylab* and *Russian MIR Station*, all LEO systems. Therefore in all LEO robotic applications there is an important aspect of interaction with crew members, as the robots involved in these missions have to perform tasks in cooperation with the astronauts; however, also autonomy is a key issue as there are tasks to be executed by the robot itself, in the attempt to help or even to stand in for the EVA crew. The other main use of Robotics is in support of scientific investigation, exploration and, in the future, in-situ resource utilization of planets. Despite the great variety of the robotic systems (e.g. aerobots, moles, hoppers) that can possibly being employed in planetary environment, present space missions are only using wheeled rovers: this is because of both the fairly traversable terrains currently being explored and the inherent simplicity of wheeled locomotion. However there are also studies about future mission scenarios where extremely rugged terrains (e.g. cliffs, craters) need to be negotiated: for these scenarios legged rovers outperform wheeled ones.

Robotics enables current missions on planetary surfaces (operations such as mobility and exploration) and on orbit applications (operations including assembly, inspection ad maintenance); moreover, it turns out to be essential for all future space exploration activities, as they aim to develop applications on both robotic autonomy and human-robot cooperation.

A good example of flight demonstrated planetary surface exploration is represented by the NASA's *Sojourner*, shown in Figure 1.1.1, which deployed from the *Mars Pathfinder* lander in July-September 1997. The only autonomous capability this rover had was that of avoiding obstacles along its path, while executing detailed command sequences determined by human controllers. Its maximum speed was 0.01 m/s and during its 83 sol mission it traveled several hundreds meters.

Figure 1.1.1. Sojourner received detailed command sequences and its autonomy was very scarce. Image credits: NASA/JPL.

The NASA's *Mars Exploration Rovers* (*MERs*), launched on the beginning of 2004 and depicted in Figure 1.1.2, surpassed Sojourner's capabilities. Even though they possessed some instrument capability and automatic navigation functionality, MER rovers too were essentially teleoperated by command sequences, with telemetry and images directly downloaded from Mars each communication opportunity and a new sequence of robot actions uploaded for the following cycle.

In contrast, terrestrial robots have demonstrated greater capabilities in exploration activities; in fact, many robotic vehicles can autonomously perform multi-kilometer traverses and even try to seek and identify meteorites. Robots such as the NASA's *K9* and *FIDO* (Figure 1.1.3) [9] [10] can approach targets and place instruments in contact with them; moreover, they have proved to be very useful in performing autonomous operations in the service of remote science teams.

These abilities demonstrate that, for the next generation of planetary rovers, it will be possible to traverse long distances and intensively and autonomously investigate sites in less time and with less human oversight. Finally, although

Figure 1.1.2. An artist's conception of a NASA Mars Exploration Rover on the surface of Mars. Image credits: NASA/JPL/Cornell University.

there are advanced rovers with capabilities for automated science data collection, none of them can be significantly involved in science data interpretation: human scientists are still those who guide the investigation and the rovers are only tools they use to achieve their exploration goals.

Hence in the future, mobility and navigation will not be any longer barriers to planetary exploration: long traverses and access, sometimes with specialized robots, to most locations on a planetary surface will be possible. Future robotic platforms will have well established navigational capability and so will be able to rove most interesting areas; development in this sense will focus on sensor fusion, in order to get information on the surrounding environment from many different sources, and on path planning, in order to find the best path in this multidimensional information space. Real-time characterization of obstacles and terrain features must be strongly developed, too, in order to achieve a higher level of safety during rover navigation.

Figure 1.1.3. K9 and FIDO during field operations. Image credits: NASA/JPL/ARC.

Moreover, robot self-awareness is an interesting matter as it concerns the capability to monitor system health and safety, acting accordingly to them; the robot must pay attention to the resources used to accomplish its mission, in particular in terms of time and power consuming.

Humans will still be present in the robot control loop, as they will be using remote surface explorers, but autonomous platforms will not need moment-to-moment commands in order to perform their tasks. Men will go on receiving and interpreting collected data, while robots will be helpful in collaboratively seeking information. Innovation will allow robots to execute complete missions and topics such as localization, perception or reasoning technologies will not preventing anymore these future systems to accomplish their objectives with high reliability over great distances. Still, fundamental limitations will be brought, as ever, only by physical properties such as power storage.

Therefore rovers will interact directly with engineers and will understand high-level goals, instead of low-level actions; however, a scientist-like skilled robot (above all in terms of observation and data interpretation) is and will continue to be a significant challenge: rovers will not have remarkable cognitive and perceptual abilities and will thus perform within narrow areas of expertise, linked to their specific mission.

Besides planetary robotics, there are a lot of in-space missions, focusing on operations such as component assembly, inspection and maintenance, typically damaged component replacement. Examples of current in-space robots are *Space Shuttle* systems and the *International Space Station* Remote Manipulator System (RMS), directly teleoperated and used only for gross component assembly. Ground testbeds, such as *Ranger* [11] or *Robonaut* [12] by NASA as well as in-space experiments like *ROTEX* [15], demonstrated more dexterous manipulation, including opening panels and connecting cables, still under teleoperation. Some other ground testbeds, such as *Skyworker* [13], were used to obtain autonomous assembly of well-suited components. Also remote in-space inspection tasks were demonstrated through the use of teleoperated robots such as the NASA's *AERCam Sprint* [14], shown in Figure 1.1.4.

Figure 1.1.4. AERCam Sprint during a flight test from the Space Shuttle cargo bay. Image credits: NASA/JSC.

Probably in the next future, robots will approach or even exceed a *space-suited* human in terms of mechanical dexterity, for example in assembly and maintenance operations, where as achieving the dexterity of a human hand unhampered by pressurized gloves is more and more challenging. Careful system engineering is required for autonomous assembly and maintenance in space as the compatibility between robots and the facility they are building must be ensured. On the other hand, automated inspection capability seems to be well established and safety reasons are the only barrier to a broader use of it.

More in details, current capabilities for space robotic assembly are represented by the Remote Manipulator System of the International Space Station. These robot arms, still teleoperated, can move large components and mate them, under a strict human supervision. Fine assembly, too, has been demonstrated by such testbeds as Robonaut, depicted in Figure 1.1.5, and Ranger.

Autonomous delicate assembly tasks will be feasible in a short time. With an intense effort, complicated structures will also be assembled in space by robots, though under supervision and, sometimes, teleoperation from space or ground-based humans. The human element removal, in this kind of operations, however, will be possible only after the development of breakthrough technologies.

For what concerns in-space inspection, there are not so many active robots at this time; as already mentioned, AERCam Sprint was an experiment of a free-flying camera, conducted in 1997, while the Germans designed a robot called *Inspector* in order to inspect MIR Station, but it failed in flight [8].

Maybe in the future, space robots will be able to perform both autonomous inspection of exterior surfaces and limited autonomous screening of sensor data. With a considerable effort, a robot will perhaps be autonomously able to inspect most exterior surfaces and detect anomalies on the Space Shuttle and ISS.

Speaking about in-space maintenance needs, both Shuttle technologies and Space Station Remote Manipulator System can move large objects but cannot execute sophisticated maintenance. Teleoperated robots doing maintenance, such as ROTEX and *ETS-VII* [16], have been employed in several in-space experiments in order to demonstrate robotic effectiveness in these operations. More dexterous robots, such as the Canada's *Special Purpose Dexterous Manipulator* (*SPDM*) [17], represented in Figure 1.1.6, can, for example, change out components under teleoperation; with some more effort, these robots will be able to autonomously access and change out obstructed components, while autonomous diagnosis and repair of arbitrary faults (that often even astronauts

Figure 1.1.5. Robonaut is a humanoid torso. It is teleoperated by a human who can see through Robonaut's eyes and control its hand positions with sensorized gloves. Image credits: NASA/JSC.

cannot achieve on orbit) need breakthroughs.

The Space Shuttle and ISS remote manipulators were used in order to move crew members from one location to another and to assist them while moving assembly components; the teleoperated robots Ranger and Robonaut performed tasks such as handing over tools and holding objects needed by the astronauts. In particular, the *European Space Agency* (*ESA*) funded a project developed by *Thales Alenia Space* and called *EWM* (*Eurobot Wet Model*), aiming to design and realize a prototype of a servicing robot for the ISS. More details about EWM will be given in the following of this Thesis.

Besides in-space robotics, great importance is given to planetary robotics: issues about robots assisting humans during surface EVA are being explored [18]. These robotic assistants have demonstrated, during field tests with suited astronauts, their ability to help humans deploy a solar panel and cables or to follow them while carrying equipment. However, a reliable and more autonomous behavior would be desirable; natural language and gesture interfaces would help

Figure 1.1.6. SPDM is an extremely advanced, highly dexterous dual-arm robot designed to attach to the end of the Space Station RMS. It is teleoperated and can carry out delicate maintenance and servicing tasks. Image credits: NASA.

these systems to be somehow limited team mates for the human crew. Furthermore, particular effort is needed in order to make the robot possess the capability to work side by side to EVA crew members: safety considerations limit the physical proximity.

To this aim, ESA is funding a project (of which Thales Alenia Space is the prime contractor) named *Eurobot Ground Prototype* (*EGP*), aiming to develop an effective robotic crew assistant that could really help astronauts during planetary missions (see [19] and [20]). More details about this project will be illustrated in the following chapters of this Thesis.

Surely mobility is a fundamental capability for space robotics; complexity dramatically increases with the desired level of autonomy and also safety plays a key role in the achievement of an effective navigation through unstructured and unknown environments. Functionalities such as localizing in the environment, collecting scientific data and navigating while avoiding obstacles found wide accomplishment but only within limited autonomy conditions; some more autonomous capabilities, such as identifying goal locations and planning paths to them, are required in order to obtain effectiveness in planetary exploration. Many significant capabilities (monitoring system state and health, reactively planning complex operations, acting in a resource-saving manner, building maps, efficiently collecting scientific data) are strictly necessary in order to achieve longer mission durations and distances, greater science return and reduced operation effort. Mechanical capabilities as well as energy are also relevant to these purposes; all these individual functionalities combine to bring about the overall achievable performance in terms of distance, duration, speed, complexity and reliability. Issues about extreme terrain mobility and remarkable science operations, for example, have been addressed by *Dante* [21] and FIDO. Planning capability and obstacle avoidance improvements will enable a significantly better exploration of unknown and hard terrain, through an on-line adaptation to the terrain being currently explored. Similarly, vision system advances will result in the ability to navigate autonomously toward features of interest in a continuous manner, while still avoiding obstacles and optimizing path control. Moreover, simultaneous localization and mapping are capabilities considered as known in theory but there are still many problems about data association. Intense development is needed in order to endow rovers with the ability to build a global map of natural terrains through the fusion of all their local sensor readings together with information provided by orbital data. A progress in the data association problem will result in making the localization of planetary rovers reliable in practice. Autonomous exploration will achieve a great level of competence when robots will be able to seek out anomalies and generate discovery plans, thus collecting relevant data autonomously. Fundamental problems, difficult to be solved without unexpected breakthroughs, are linked to physical limitations: performances, for example, are still tied and limited by available power and technology, as well as mission duration, governed by it. Moreover, also the ability to recover after a failure represents a crucial aspect but is not easy to be gained by robots. Exploration is also held back by the different terrains it has to deal with. In fact, a wide variety of terrains have been considered, from soft dune ones to boulder fields, till hard steep slopes, and almost for any of them a specialized rover has been designed, and sometimes even realized and tested. However, some terrain elements, such as cliff faces, still represent big challenges and prevent these platforms to succeed at traversal. Rovers able to perform their tasks in a completely autonomous manner are a step toward truly independent rovers that could deduce their actions directly from mission objectives.

Operations like surface instrument deployment were performed by rovers (such as MER), using supervised autonomy and only within several cycles of actions commanded by humans; to this end, one of the desired objectives of space robotics is that of developing robust, highly autonomous sample collection and instrument placement capabilities, which were already demonstrated by terrestrial robots (while space robots are far behind with respect to them). For these reasons, EGP is a fundamental step in the field of space planetary exploration in order for robotics to prove itself as able to realize effective assistants to men.

Another important robotic ability is science planning and perception; there are rovers with on board planners that can maintain lists of goals with different priorities and several constraints among them. Moreover these platforms must be able to execute fully autonomous operations, at least in relatively simple environments. They should be endowed with the ability to seek anomalies by themselves and to generate discovery plans, thus collecting interesting scientific data and making human effort decrease. Even if there are many studies in this field, the development of robotic performance at the level of a human scientist is still a big challenge.

There are basically four different aspects that are to be regarded as fundamental in current and future space mission preparation: robustness, whole system design, mission competence and virtual presence.

Robustness means the property of a system to go on working even if faults, anomalies or unexpected conditions occur; this feature achievement is a relevant challenge for engineers because robots must be able to react to several stimuli from the external environment and, often, some situations cannot be neither planned in advance nor approached with standard verification and validation techniques. Robotic systems, usually, are weak in many different regards and they cannot surely reach human adaptability. Robustness can be achieved by realizing rovers that could address faults through self-diagnosis and recovery and that could live a long time in spite of issues such as power, temperature or stability. Attaining human intelligence and flexibility would be a great step toward robustness and effectiveness during missions too; this gap can be partially filled in by teleoperation or through advice given to the robot whenever it encounters problems it cannot deal with all by itself. Anyway, as few capable robots have been launched in space, it is difficult to validate their behavior with respect to the space environment; hence, for the experimental nature of robotics itself, without suitable experimentation, it is very difficult to achieve a high level of robustness.

Key issue in the preparation of a space mission is the whole system design: robots neither can work in isolation nor are effective while working into a system that was not designed for them. In fact, when designing a mission, scientists consider overall system, that is not only the robot but also supporting infrastructures (power, communications, navigation and maintenance) and the human component. This is far more important to the success of robotics than any single robot-specific capability such as mobility, dexterity or intelligence; an appropriate system engineering can greatly increase the robustness of robotic missions.

In spite of this research for autonomy and high reliability, humans will always be involved in space exploration, either as consumers of the data collected by robots or as directors of robotic activities. The point is to let robots work by themselves as much as possible: the human must concentrate on mission objectives and scientific strategies and must not be forced to coordinate the minute-to-minute activities of the automated system. Usually, space robots work with simple primitives like "put your manipulator in this specific configuration" or "go to this exact place"; humans string together these actions in order to accomplish mission objectives. A fundamental ability for robots is to process more complex goals such as "explore that area and report something interesting" through improved cognitive capabilities including adaptation, planning and diagnosis, without any a-priori knowledge of the task to be accomplished.

Finally, human virtual presence is a long-range goal for space robotics; it is about allowing for human cognitive presence in space or on a planetary surface without any actual human physical presence. This, for instance, is the case of an engineer overseeing the assembly of a complex space telescope or troubleshooting it while comfortably sitting in his chair; or of a planetary geologist roaming Mars, picking up rocks and examining them without leaving his lab on Earth. This kind of technology is very powerful and cannot disregard issues about high-bandwidth and low-latency communications; nevertheless robotics too must contribute to this purpose by replicating both the dexterity and the sensing modalities of a human being.

In conclusion, the main features for an effective space robot are:

- dexterous assembly: dynamic control, assembly planning and high level of autonomy are essential in order to let the robot build anything and are to be strongly experienced;

- human/robot collaboration: cooperative skills come together with the capability to understand natural language and human gesture, in order to effectively help astronauts and cooperate with them;
- planetary surface access: localization, mobility planning and reliable terrain perception are required in order to explore every kind of surface;
- comprehensive surface investigation: multi-modal sensing skills, science models and evaluation are fundamental in order to conduct efficient scientific research and useful data collection.

Eurobot Ground Prototype project was conducted having these main aspects in mind; EGP, in fact, is intended to demonstrate abilities like dexterous manipulation, cooperation with astronauts and a high level of autonomy during its operation execution.

For what concerns the Eurobot Ground Prototype, Thales Alenia Space is the prime contractor of the project, while the European Space Agency is the customer and founded the project; part of the author's work on EGP was done at Thales Alenia Space - Italy premises, in Turin.

1.2. Thesis Overview

Surely, with the aim to achieve such a desirable level of both autonomy and cooperation with human beings, research on suitable coordination algorithms and vision are fundamental in order to assure the mission completion with a high level of safety and reliability. Moving from these considerations, this work is organized as follows. Chapter 2 describes the overall ESA Eurobot Programme and, in particular, some remarks on both EWM robotic structure and its vision system are emphasized. The following two Chapters describe in details EGP project objectives and innovations in the context of space exploration. Chapter 4 fully illustrates EGP control algorithms, where as Chapter 5 shows the research results obtained within this project. Different EGP operations are described, above all two tasks demonstrating the autonomy and cooperation skills accomplished by this robot. In this Chapter, great importance is given to vision and to its fundamental role in EGP operation success. Moreover, Chapter 6 provides details about stereovision and its effects on space missions, both in terms of robotic efficiency with respect to mission goals and for purposes not closely related to robotics, but equally important for space exploration. Finally, some considerations about the work in progress on this field are provided in Chapter 7, as well as some more general conclusions and indications on future possible research activities.

Chapter 2

Eurobot Wet Model Description

2.1. Project Objectives

The ESA Eurobot experience originates from the future space exploration plans that include human missions to the Moon and Mars. In 2003 the European Space Agency decided to launch a new robotic programme, in preparation for human exploration; more in details, they resolved to investigate the employment of an EVA assistant robot that could support the crew during its EVA sortie. That assistant, named Eurobot and shown in Figure 2.1.1, uses a novel approach to space servicing robotics in human-compatible infrastructures such as planetary orbiters, planet surface bases and, above all, ISS (see [47] and [48]).

ESA astronauts took part in Eurobot design and thus helped engineers to define the system concept; this collaboration led to the realization of a robot, whose force is comparable to that of a human, equipped with three arms (one of which has the role of a sort of "leg") mounted on a central torso and a head with stereo vision capabilities for scene overview. In order to obtain an early feedback from non-robotic skilled people on both operations and interaction aspects, a consortium led by Thales Alenia Space - Italy has developed for ESA an "underwater" model for Eurobot, called *Eurobot WET* (*Weightless Environmental Test*) *Model*, which is capable of operating in a Neutral Buoyancy Facility (NBF). In fact EWM was tested in a pool in such a way to better simulate both a space environment (weightless) and a realistic scenario (proper infrastructures and divers acting as astronauts during tests). EWM whole system is made up of essentially two different part:

- *Dry Segment*: the part of the system that is outside of the pool. Its main components are the power supply and the robotic control station, useful to control the system. Its human operator plays the role of a crew member on the ISS using an Intra Vehicular Activity (IVA) workstation;

- *Wet Segment*: the part of the system that stays underwater (also called underwater robot or mobile system).

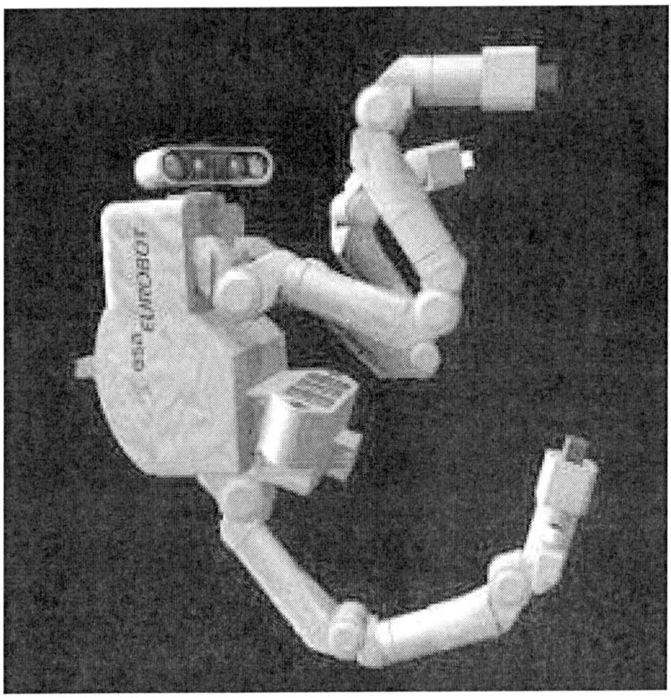

Figure 2.1.1. Eurobot Mobile Segment Overview. Courtesy of ESA.

The development of Eurobot Wet Model has been carried out according to the main following principles: mechanical, thermal and electrical design must differ substantially from the Eurobot Flight Model, taking into account the different requirements related to the particular wet environment and the reuse of commercial and off-the shelf components (end-effectors, cameras and electronics), for cost and schedule reasons. Moreover, as exchanging an end-effector in the underwater environment would be a very considerable effort, this very desirable requirement has been abandoned, thus imposing some limitations to the possible tasks to be executed by the WET model. Collision avoidance with respect to moving or not foreseen obstacles in an environment like the ISS was not exhaustively considered too: Eurobot will be a testbed for future experimentation and implementation. A set of possible demo operations was proposed, recalling the limitations mentioned above, and a test program to be performed at ESA Astronaut Centre (EAC) was scheduled.

EWM was thus realized to demonstrate the overall concept of some of the operations for which a robotic intervention is desired. Differences, mainly induced by the test environment, were envisaged between the Eurobot Flight Model and Demonstrator (EFM/EFD) and EWM; in fact, as already said, no end-effector exchange was planned to be part of the EWM operations. Only a subset of the EFM tasks was demonstrated by the EWM, so it was possible to follow design guidelines that were only dedicated to the tasks to be performed in the pool. Also safety and failure tolerance requirements were more relaxed with respect to EFM, which is conceived as an upgrade of EWM (even if safety issues were considered in order to allow divers to work closer to EWM). Figure 2.1.2 points out the differences between EFM and EWM: the most evident one is the size of EWM components (on the right). In fact, as it had to work underwater, its electronics had to be protected and its weight lightened by floating elements, in such a way to reproduce an as much as weightless scenario.

As previously mentioned, EFM was conceived to be employed on board the International Space Station. Once completed, the ISS will be made up of many large elements, some of which pressurized; it will be the largest man-made infrastructure orbiting around the Earth and will need both preventive and corrective maintenance interventions in order to remain operative and keep flying in next years. With the purpose of allowing its in-orbit maintenance, hundreds of Orbital Replaceable Units, namely different kind of equipment, are externally mounted onto the ISS; these elements can be replaced either by a human while in Extra Vehicular Activity or by a robotic assistant. On the US segment alone, the assembly of more than 300 ORUs is scheduled, from small items such as gyros to large, heavy batteries.

Currently on board the International Space Station, crew member presence is very scarce: it is now limited to 2-3 astronauts and will be extended up to no more than 6. For this reason, it is fundamental to optimize crew member time usage: they are required for tasks needing human intelligence, such as scientific experiments or troubleshooting activities, for time critical decision making and

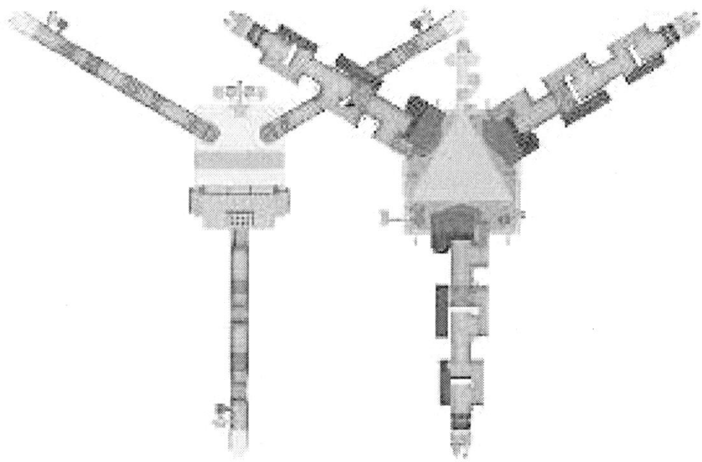

Figure 2.1.2. Eurobot Flight and Wet Model side by side. Courtesy of ESA.

very high level dexterous operations. All other activities, which are routine or well defined and controlled, can be entrusted to automation and highly reliable robotic systems.

In order to achieve such a purpose, robotics must focus its effort on the realization of better systems to support the Extra Vehicular Activity performed by the crew: the logistic tasks are, in fact, well executed by the current ISS robotic capabilities. Moreover, this kind of EVA sortie (i.e. spacewalk), besides being dangerous, wastes a significant amount of human resources because of the preparation time required for the activity and also of the rest needed by the crew after the EVA execution. In this sense, an on board assistant would relieve astronauts of many spacewalks, even maximizing the number of tasks carried out during a unique EVA sortie, thus reducing the required number of EVAs.

The International Space Station requires this kind of operations in order to support several activities, including the upcoming ISS assembly missions, maintenance, servicing, and repair of the ISS and other large space structures; moreover also maintenance, servicing, and repair of ISS payloads and experiments, removal and replacement of Orbital Replaceable Units (ORU) and contingency operations are needed.

A robotic system complementary in capability to an EVA crew member would be a great benefit for all these aspects of ISS extra vehicular operations.

Such a system would decrease the workload on an EVA crew member, thereby increasing the overall safety of the ISS crew and saving both valuable consumables (O_2 and N_2) and, as previously mentioned, crew time. The fulfillment of these support needs is objective of EWM, a robotic system comparable in capability, including manipulation skills, to a human.

For planetary missions (Moon, Mars) it is equally important to have a robotic system to do both EVA-type work when no human crew is being flown yet and cooperative work in a later stage at crew presence. In this context EWM was exploited as a starting point in order to design, build and operate a working robotic demonstrator devoted to planetary missions (such a prototype is EGP).

On the other hand, the robotic system demonstrator known as Eurobot Wet Model had to remotely perform some comparable EVA tasks while controlled by a remote operator outside of the pool, similarly to a crew member in the ISS using an Intra Vehicular Activity (IVA) workstation. EWM had to work underwater within the EAC (European Astronaut Centre) mock-up, simulating a Space Station structure utilizing standard EVA interfaces. With its standard end effectors and a tool, Eurobot had to perform or support routine EVA tasks like the inspection of the Space Station, installation or removal of a small dummy payload, illumination of the work site as a support to astronauts in EVA.

2.2. The International Space Station

The International Space Station, being assembled in Low Earth Orbit, is an international research facility: operations on it are performed by people from the five space agencies that participate in the programme. These agencies are ESA, NASA, JAXA (Japanese), CSA (Canadian) and RKA (Russian). The station, shown in Figure 2.2.1, is, up to 2010, the largest artificial satellite and its completion is scheduled by 2011.

On board the ISS crew members conduct experiments regarding physics, biology, human biology, astronomy and meteorology; this particular research facility has a microgravity environment to allow these experiment execution. Many space station projects are joined together to create ISS global programme: from the American Freedom, to the Russian Mir-2, the European Columbus and the Japanese Kibo, that was launched on July 2009 by American space shuttle. For details about ISS and conducted experiments refer to [45] and [46].

ISS advantage over space shuttles and other spacecrafts is that of providing a long term presence in space environment: extended studies can be easily

Figure 2.2.1. The International Space Station on 25 November 2009 as captured from the crew of the departing Space Shuttle Atlantis during STS-129. Image credits: NASA.

conducted, exploiting the unusual conditions on board the station. Moreover, thanks to the permanent crew presence, scientists on Earth can access on board achieved data and eventually change experiments, even launching new ones. Essential studies on human life in space are being conducted, above all in order to improve current knowledge about human body reaction to long term space permanence; these experiments are aimed to evaluate the effects of long interplanetary cruises on astronauts (for instance in anticipation of the six months required for a journey to Mars). Furthermore, as mission risks are reduced on ISS, a lot of experience on systems on orbit can be acquired and many tests on spacecraft systems can be performed; this is very important, too, for the next scheduled long duration Moon and Mars missions.

The complete ISS, whose main components are shown in Figure 2.2.2, is designed to be made up of 16 pressurized modules (11 of which were already launched) including laboratories, living quarters, docking compartments and so on. Europe developed, for instance, *Columbus* (whose design was conducted by Alenia Space), devoted to biology, biomedical research and fluid physics. European *Cupola* and *Node 3*, or *Tranquility*, too, are scheduled to be launched by a space shuttle on February 2010. Cupola, as suggested by its name, is an observatory module with the aim of providing crew members with a direct view of both docked spacecrafts and operations executed by robotic systems on board the station. On the other hand, Tranquility is designed to provide advanced life support functionalities and to supply the station with four more berthing locations that can be used for space vehicle docking or for new pressurized modules. Finally, one of the last assembled modules, *Kibo* (or *JEM - Japanese Experiment Module*), is devoted to several research fields, among which communications, Earth observations and biotechnology.

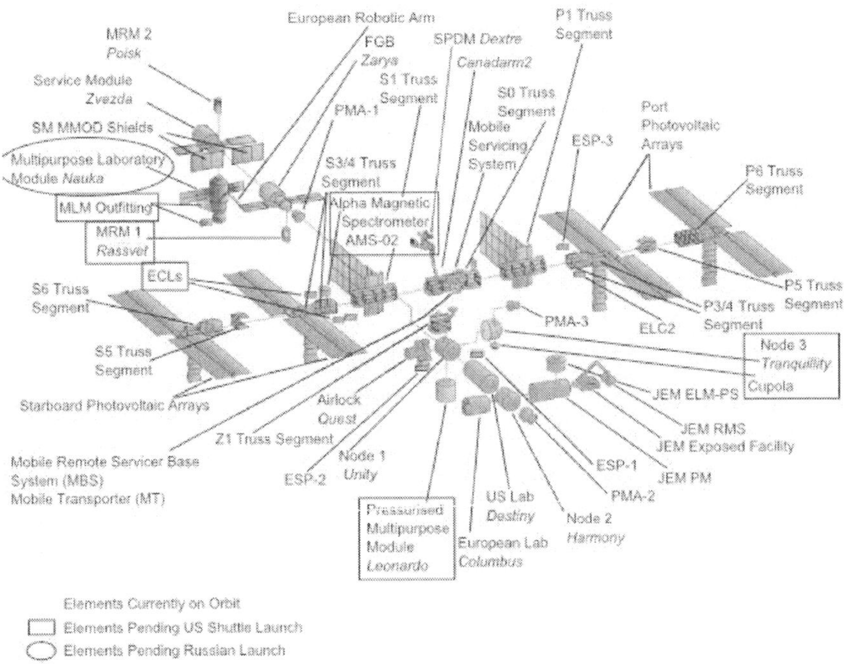

Figure 2.2.2. ISS modules as of November 2009. Image credits: NASA.

2.3. EWM Tasks

EWM had to perform compound tasks, defined as sequences of simpler actions; this made the mission definition more flexible, allowing the user to test and redefine complex tasks whenever necessary.

Eurobot main tasks, detailed in the following, were tested and are:

- Walking
- Handover of tools to/from astronauts

- Small equipment replacement
- Inspection

2.3.1. Walking

One of the most important ability to be achieved by robotics in space missions surely consists in being able to efficiently support astronauts and even replace them in dangerous or difficult tasks, saving their valuable time. EVA assistant robots are useful before, during and after the EVA sortie because they can

- do initial preparatory work, such as close-up inspection, at the intended work site;
- transfer tools and equipment needed by astronauts;
- assist humans during the work itself, providing addition support, in terms of viewing and lighting capabilities or powering tools too;
- clean the work site, store away tools and equipment when the work has been completed.

Just in order to gain these capabilities, EWM walking task accomplishment was judged very valuable. This task may be split in actions such as "grasp/ungrasp", "move step" and "turn body". For what concerns the first action, it can be executed both automatically and with interactive human control, in some cases when it is somehow difficult for the robot to identify EVA crew interfaces (yellow handrails). Some frames from grasping simulation are depicted in Figure 2.3.1. The moving action consists in a coordinated motion in which one arm is grasping an handrail and a second arm is left free to move, in order to reach the pre-grasp position of another handrail, as can be seen from Figure 2.3.2.

Figure 2.3.3 shows the "turn body" action: position and orientation of the robot body are modified in order to allow the robot itself to reach next handrail to be grasped. Finally, some frames taken from a simulation illustrate EWM walking in Figure 2.3.4, while Figure 2.3.5 shows EWM performing its walking task along an ISS-like structure realized for underwater tests.

(a) Starting position.

(b) Pre-grasp position (with errors).

(c) Aligned position (after vision assistance.)

(d) Grasped handrail.

Figure 2.3.1. EWM handrail grasping task. Courtesy of ESA.

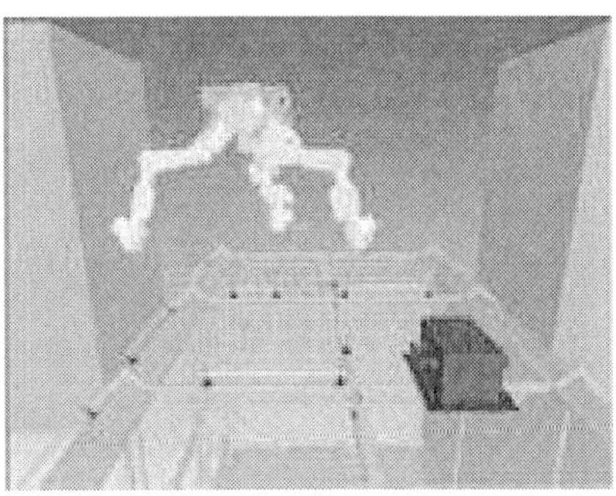

(a) Initial position.

(b) Pre-grasp position (with errors).

Figure 2.3.2. EWM move step task. Courtesy of ESA.

(a) Initial position.

(b) Final position with body oriented in handrail direction.

Figure 2.3.3. EWM turn body task. Courtesy of ESA.

2.3.2. Handover of Tools to/from Astronauts

This operation was required essentially for the handover of the special screwdriver tool needed by EWM in order to perform its screwdriver-like operation; however this tool exchange turned out to be too difficult for the robot. In fact, the handover of objects with an astronaut (a diver during pool testing) required a human presence very close to the system; this action thus would have needed a very careful analysis, above all about safety issues. Moreover, reception of a tool from an astronaut or from a swimming diver would have meant effective visual tracking and dynamic following of the target.

Therefore the handover of this tool was rethought in order to be implemented, in a quite passive mode, with the following phases:

- the robot opens its end-effector;
- the astronaut/diver puts the tool in the robotic gripper;
- the robot, under operator control, gently closes its end-effector;
- the tool is accommodated by the robot in a proper place onto its body;
- the tool is taken by the robot from its own body whenever it is needed.

In order to achieve this behavior, some of the involved actions are "get tool" and "store tool", shown in Figure 2.3.6 and Figure 2.3.7.

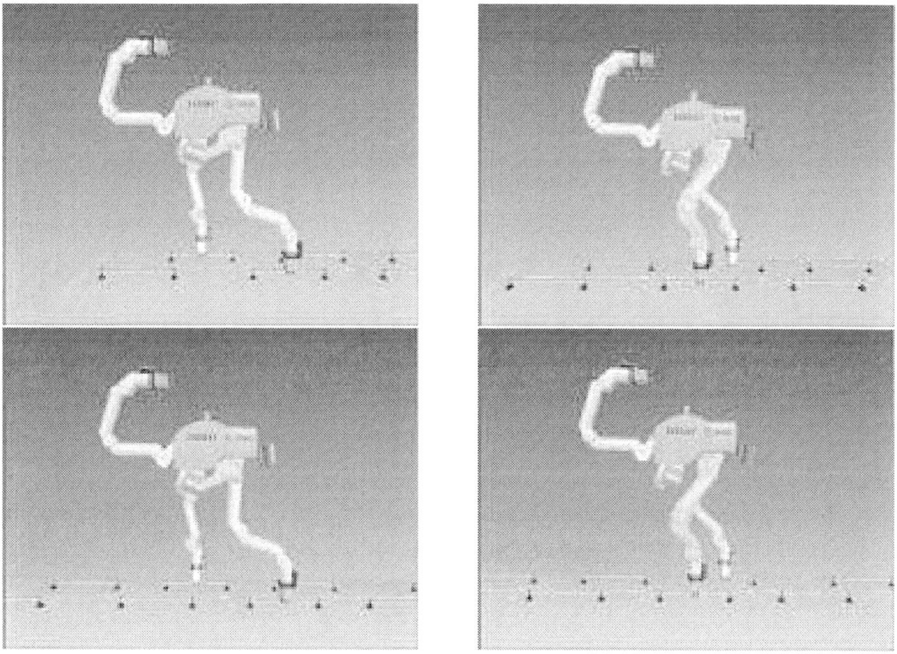

Figure 2.3.4. EWM walking. Courtesy of ESA.

2.3.3. Small Equipment Replacement

Besides its support to EVA sortie, Eurobot had to be able to perform also small equipment replacement in an automatic mode. Though ISS is already equipped with systems that can perform robotic exchange of equipment, Eurobot can be a valid complement to the existing infrastructure.

Equipment replacement is accomplished through sequences of actions such as "grasp/ungrasp ORU", "remove ORU", "move step" and "install ORU" (usually in a different place). Figure 2.3.8 shows some frames taken from a simulation of the "remove ORU" action.

2.3.4. Inspection

Currently, external robots on ISS provide eyes for the orbital station crew (the *Canadarm2* provides camera view to the station exterior). Both on board crew and ground operators highly appreciate the robotic capability to perform close-

Figure 2.3.5. EWM walking on the ISS-like structure in the pool. Courtesy of ESA.

up inspection with high resolution; for this reason, the capability to access ISS areas otherwise difficult to reach (e.g. docking ports) was foreseen for Eurobot. An example of this ability is the removal of MLIs or debris shields, in order to inspect them and evaluate station conditions.

2.3.5. Eurobot Wet Model Overview

The objective of Eurobot development program was to realize a robot for ISS servicing; this robot had also to constitute the base for future planetary missions (landing on the Moon or on Mars) and for other future human spaceflight scenarios. Within this program, ESA preliminarily studied a semi-autonomous robotic system, which was intended to offer the ability to

- manipulate payloads or parts of the space station infrastructure;
- carry and transport ORUs, tools, and other equipment for crew members during EVA;

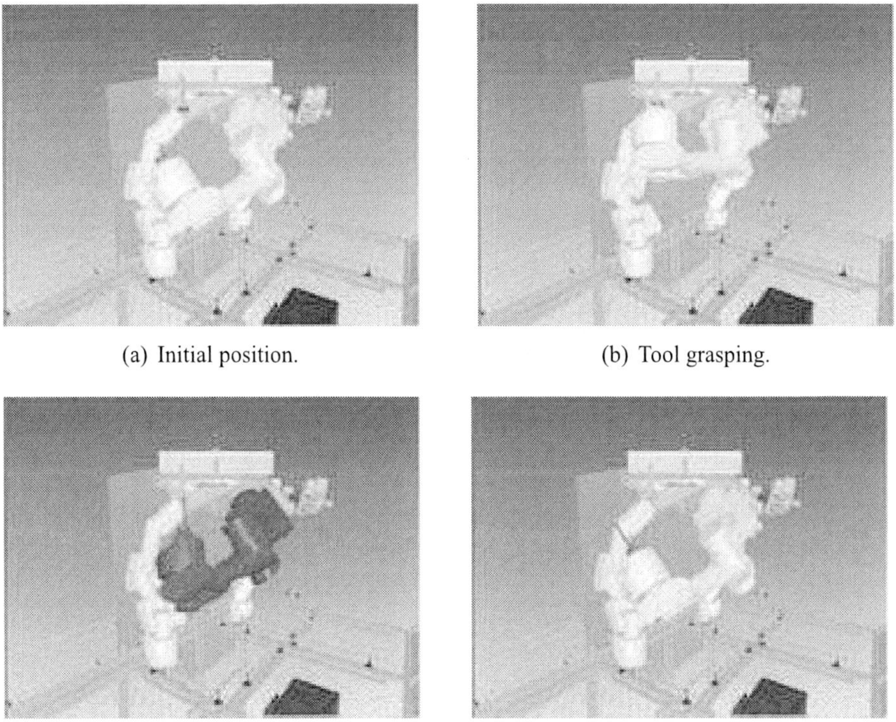

(a) Initial position. (b) Tool grasping.

(c) Tool extraction. (d) Final safe position.

Figure 2.3.6. EWM "get tool" task. Courtesy of ESA.

- perform equipment/surface inspections.

On the basis of this study, Eurobot project was started, covering a feasibility study and a development of Eurobot model. Figure 2.3.9 shows the concepts of the robot configurations prepared for the Preliminary Requirements Review (PRR), from which further evaluation on final constraints was carried out.

EWM had to be similar and not to exceed the dimensions given for the flight model; the requirements for EFM dimensions were driven by on orbit scenarios (e.g. possible need to pass across hatches). These requirements were not applicable to EWM, whose dimension had to be constrained only by the operational environment. In addition, EWM size was also influenced by the specific protections and design solutions to be implemented to allow underwater operations.

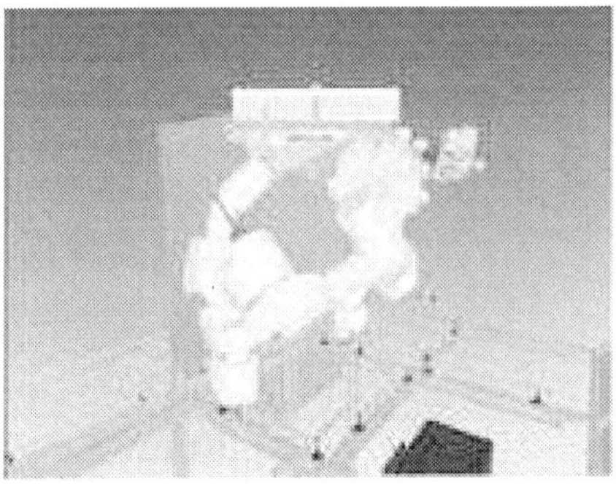

(a) Initial position.

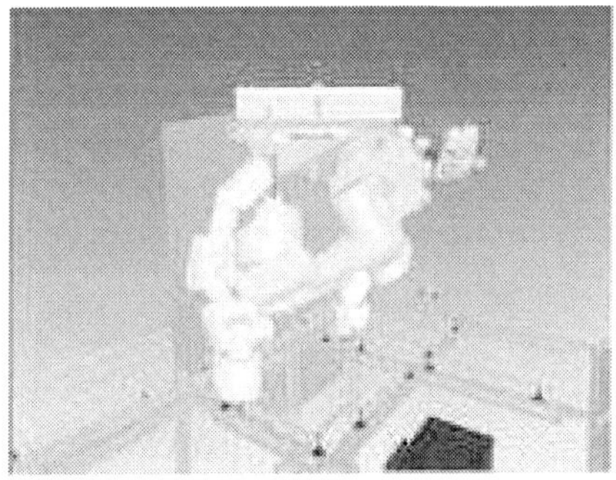

(b) Tool grasping.

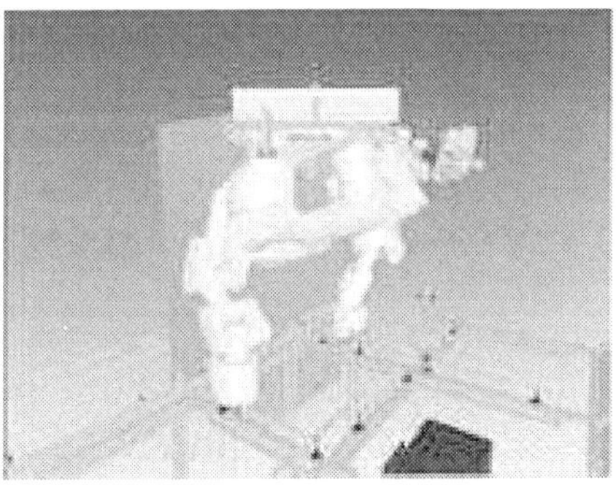

(c) Tool extraction.

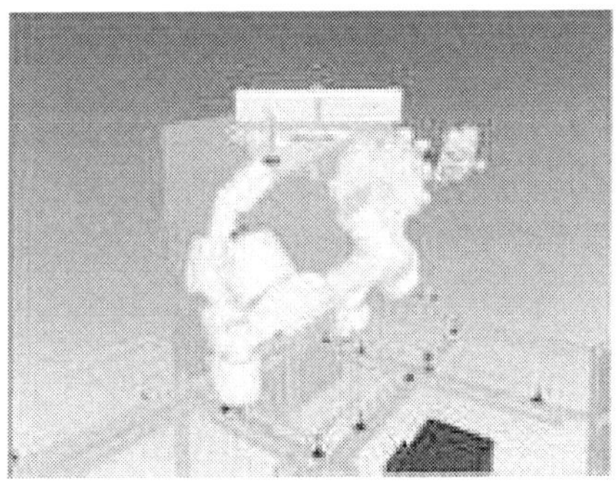

(d) Final safe position.

Figure 2.3.7. EWM "store tool" task. Courtesy of ESA.

(a) Starting position.

(b) Final position.

Figure 2.3.8. EWM ORU removal task. Courtesy of ESA.

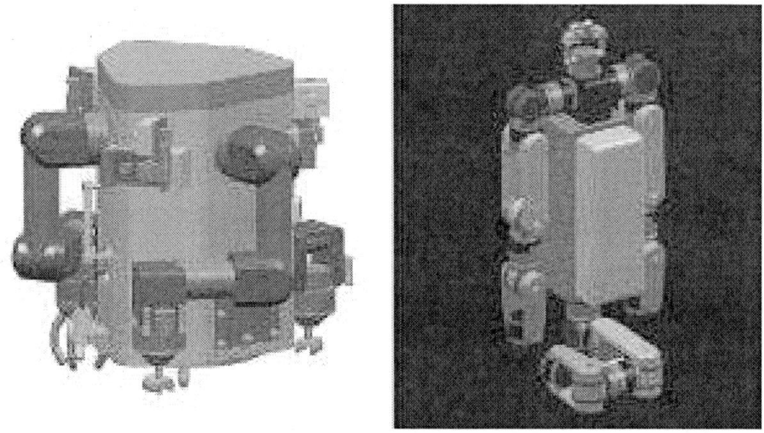

Figure 2.3.9. Baseline and optional concepts for Eurobot configuration. Courtesy of ESA.

The following Figure 2.3.10 shows the principal system architecture of the whole EWM and the interfaces between the dry segment (gray area) and the wet segment (all connections and interfaces are considered as bidirectional for commanding and feedback purposes).

Eurobot Dry Segment

The design of the dry segment consists in the definition of the system portion devoted to both human supervision of the demonstration in the NBF and execution of the activities to be performed by the ground control centre, or by the astronauts from inside the ISS during the flight mission. The dry segment includes a work-station with the Man-Machine Interface (MMI) software and the simulator that can guarantee the correctness of a given operation sequence before the robot itself is activated. In addition EWM dry segment includes also the portion of the robot hardware and software that in the nominal flight version resides inside the robot. This control station works as a simulator of the robot computer enabling the motion and vision functionality required by the demonstration.

The main components of the dry segment are:

- *Robotic control station*
  The robotic control station allows preparation, control and monitoring of missions, tasks and actions. These operations can be performed either

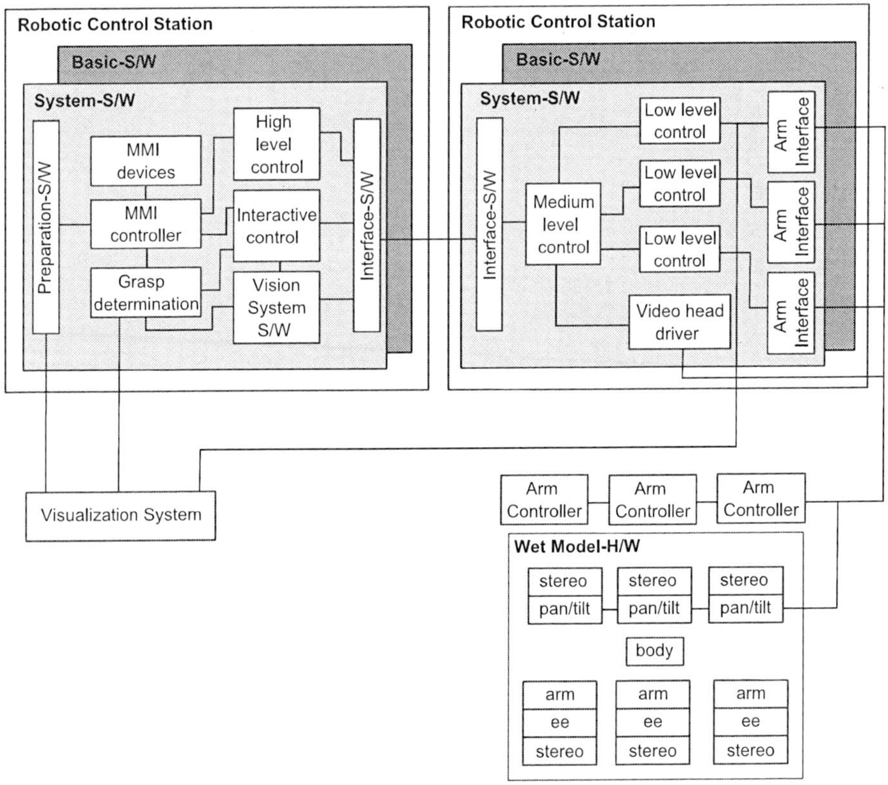

Figure 2.3.10. Top level architecture of the Eurobot Wet Model. Courtesy of ESA.

automatically or interactively under operator control. Eurobot hardware consists of one or two workstations. Some of the main functionalities are the man-machine interfaces, the grasping point determination, the high-level planning and the interactive control. Especially the grasp point determination and the interactive control are based upon the vision system, which will be described in the following of this chapter.

- *Guide Navigation Control Station*
  The Guide Navigation Control (GNC) station has to perform all the functionalities in charge of the mobile robot control; in order to reduce the costs and to simplify the integration, it was decided not to accommodate the control electronics into the body of the mobile segment, but to place

it within the dry segment as a separate module.

- *Visualization System*
  Closely connected with the Robotic Control Station, it consists of a commercial robotic visualization tool and can display robot movements, through data directly received from the low-level control of the GNC. When using the control system in off-line mode, it is possible to simulate the robot behavior; in this way pre-planned path and operation verification can be performed.

Within the RCS, great importance is given to the *Vision Processing Unit* (*VPU*), that controls all Eurobot camera streams and processes them in order to accomplish all the vision-based tasks.

Eurobot Wet Segment

The Wet Segment, as previously said, is the subset of Eurobot components that worked inside the pool during test campaigns; it consists of a waterproof central body, including all the integrated electronic hardware, three waterproof arms (each of which has 7 d.o.f.), endowed with as many end-effectors, essential to grasp handrails. Moreover, the whole vision system is part of the wet segment; it is made up of a stereo vision assembly (a kind of "robotic head") consisting of three body-mounted cameras placed on a pan/tilt unit (PTU), and one camera for each arm wrist, in order to follow more closely all operations being performed.

EWM vision system is made up of six firewire color cameras, having a maximum resolution of 1024×768 pixels and able to perform at 30 frames per second individually, or at a maximum data rate of 7.5 fps when three of them are acquired at the same time. Finally, the robot, shown in Figure 2.3.11, is equipped with JR3 force/torque sensors, one for each end-effector, in order to achieve compliant motion.

2.3.6. EWM Vision System

In such a way to make the robot be able to effectively support EVA, the development of a "smart" vision system is fundamental in order to give the robot a good perception capability, thus allowing it to interact with humans and objects in the surrounding space environment. The main functionalities for which the vision system is important are:

Figure 2.3.11. EWM performing EVA assistant mode. Courtesy of ESA.

- telepresence;
- semi-automatic or automatic target identification and grasping;
- monitoring.

In the telepresence, semi-automatic and automatic modes the images of the vision system are directly sent to the control station to be automatically processed and/or visualized by the operator. The body cameras of EWM are dedicated to the telepresence mode for simultaneously observing both the robot (especially its arms) and the surrounding environment. The body-placed pan/tilt stereo camera unit delivers stereo images that can be visualized by means of a dedicated stereo display. The cameras placed on the wrist of each EWM arm are dedicated to the semi-automatic or automatic identification and grasping of the objects to be handled.

In order to support the foreseen EWM operational modes, the vision system must be able to support different operations such as:

- moving along the ISS-like structure by the means of standard translation aids for astronauts (EVA handrails);

- carrying and manipulating payloads;
- handing over and taking tools to/from astronauts;
- performing screwdriver-like operations.

The vision system works together with the GNC software: this last, in fact, must manage all the operations regarding interactive and automatic tasks, compliant control and safety check. In particular, during automatic grasping, the vision system must support GNC station by identifying the object to be handled and its position. The VPU software duties can be summarized in four main steps:

- image acquisition (up to 3 cameras at the same time);
- camera status and parameter control;
- whenever required by the GNC, try to identify the object to be grasped and compute its position;
- if the previous phase succeeds, send the object pose to the GNC. These data are essential in order to control the robot to accomplish its grasping task.

Thanks to the firewire link, the head camera images, as already mentioned, can be simultaneously acquired and so the VPU uses two of them just as a stereo rig; these stereo images can be also displayed on a dedicated monitor that is part of the RCS and allows a stereoscopic 3D visualization, useful for human operators that have to monitor the correct operation execution. The pan/tilt unit supporting the head cameras is required in order to move them during EWM desired tasks and a simple lighting system (made up of a waterproof halogen lamp) helps the VPU to better perform image processing.

During EWM development and testing, two different automatic algorithms for object identification were evaluated: one based on the knowledge of a VRML model of the item to be handled, the other based on its shape, size and color.

For what concerns the first one, this approach was tried just because it is a general method to match any three-dimensional object, provided that its 3D model is known. The steps for this algorithm are the following:

- image processing and extraction of interesting contours;
- wireframe rendering of the object 3D model through VRML files. All hidden lines, thanks to the knowledge of the current estimated point of view, are removed;
- matching of extracted image features to model edges (projected onto the image plane);
- computation of an error function evaluating the mismatch between desired and current views;
- minimization of the above error function, in order to update the estimation of the point of view.

Nevertheless, this approach did not bring about the expected results, as it was not at all robust and reliable in the handrail identification. Hence, a new algorithm was required in order to try to accomplish the automatic handrail grasping. This new method is not as general as the previous one, in fact it is explicitly based on the knowledge of some handrail properties, such as its color. This piece of information allows to better identify the handle in the scene, but, in this way, the algorithm should be implemented again if EWM tasks required the grasping of a different object. However, with this new algorithm too, some problems still remained: sometimes the procedure failed to recognize the handrail and could not grasp it. Therefore an interactive tool was added to the robotic control station, in such a way to allow a human operator to support the vision system. Within this operative mode, in fact, the operator was able to assist EWM in the handrail pose estimation, through a smart selection of the features corresponding to the handrail in the cluttered and noisy acquired image. The involvement of a human operator is motivated by the high reliability and safety that the system must achieve with respect to its task accomplishment and by the failure chance shown by the algorithm.

2.3.7. Testing Phase and Conclusions

Many different tests were carried out, as stressed by Figures 2.3.12 and 2.3.13, both at ALTEC facility, in Turin, and at the European Astronaut Centre located in Cologne, using the different vision algorithms. The testing phase pointed out that these last were not so reliable in handrail recognition during the grasping task; this was due to the following causes:

- insufficient size of target object in the acquired image: if the handrail area is too small with respect to the whole image, algorithm convergence problems and scarce precision issues occur;
- extreme proximity: the closer the robotic arm is to the handrail, the poorer acquired images are, thus making more difficult the object identification process;
- rotation around the handrail axis (in the sense of its maximum length): it is very hard to distinguish two different poses characterized by small rotations along this axis and so the grasping becomes much less safe and reliable;
- bad illumination: poor lighting conditions can highly affect image quality, just making the automatic vision procedure fail;
- presence of a metallic grid: the grid lines of the metallic structure supporting handrails underwater sometimes got confused with the correct handrail edges. This happened when using the VRML-based method.

Figure 2.3.12. EWM during test campaign at ALTEC facility in Turin. Courtesy of ESA.

Figure 2.3.13. EWM during tests at EAC facility, Cologne. Courtesy of ESA.

For these reasons, an interactive mode for handrail recognition has been designed, and during the EAC testing this involvement of a human operator proved to be effective enough. A manual control for Eurobot arm pose adjustment was added too. This was done through the aid of a haptic device that could provide both position commands to the robot and force feedback to the operator, in order to manage both free and contact motion situations.

Furthermore, for what concerns EWM control, suitable coordination algorithms were studied and developed at Graal Lab, in Genoa, and are reported in [34]. However, this cited technique is not as general as EGP control strategy: on the contrary, it is a particular case of the last one. This new general coordination and control method will thus be detailed in the following chapters.

In conclusion, one of the most challenging part of this project was the development of both a pretty smart vision system (that could accomplish the desired vision-based tasks) and, at the same time, a suitable coordination and control architecture. Still, as these purposes were only partially achieved (sometimes the vision system could not identify the handrail location) the need for more robust and reliable algorithms arises, in order to make robotic EVA operations, such as walking along external ISS surface, actually autonomous. Finally, the generalization of the control technique is desirable for different reasons: first of all the strategy presented in [34] is sub-optimal with respect to EGP one; moreover, thanks to its general character, this new control strategy can be extended to any system, regardless of its specific robotic structure.

Chapter 3

Eurobot Ground Prototype Description

3.1. Project Objectives

As previously mentioned, ESA studied, in the frame of Eurobot programme, EWM system, a robot assistant to manned EVA sorties. EWM was equipped with three arms, each of which had a camera mounted on it, and a head stereo assembly. Designed to service the International Space Station, Eurobot system was studied up to the development of a demonstration system and a functional prototype was tested in order to evaluate its capabilities during Extra Vehicular Activities and for equipment exchange modes; this enabled to gain valuable experiences on involved challenges and potential solutions. While the first Eurobot system was studied and tested for an ISS-like environment, there is an arising need, in the frame of future space missions, to study and test Eurobot derived technologies in support to human exploration beyond ISS. Therefore, in such a way to demonstrate the possibility to use robotic technologies for exploration, a new phase of Eurobot programme determined to design and develop a ground robotic prototype, establishing the most relevant robotics functions referring to exploration missions and identifying critical areas and possible solutions to be further analyzed.

This system, named *Eurobot Ground Prototype* (*EGP*), is a mobile robotic assistant to be employed in surface operations, in order to accomplish robotic tasks in several phases of space missions. The main purposes of this demonstrator development are achievable through an extensive testing of the system and consist in the evaluation of:

- the proposed concept;
- the required functions;
- the desired performances.

Moreover, EGP realization was intended also to collect valuable feedbacks to be implemented in the prototype itself and/or in the flight project evolution.

These objectives can be achieved through several actions such as:

- expand Eurobot purpose beyond its use on ISS by demonstrating both on surface robotic tasks in preparation to human arrival and human-robot joint tasks during human presence;
- adapt and complement the existing Eurobot Model with functions such as mobility, command and monitoring, use of additional tools and robotic hands, as well as enhanced vision, autonomy and cooperative skills;
- evaluate, in a high representative scenario, the Eurobot extension to exploration;
- perform complete system tests and final demonstration of the upgraded Eurobot with its new functionalities.

3.2. Demonstration Scenarios

In support to exploration activities, robotic platforms can work in assistance to the human essentially in three ways:

- realizing the goal of a task otherwise unfeasible by a human being;
- working efficiently (i.e. not requiring full attention/control);
- sharing goals (working together).

To achieve a task otherwise unfeasible, the robotic platform shall have improved capabilities with respect to human capabilities in the considered context, for example, supplying missing capability of a person temporarily hindered by the environment (e.g. suited astronaut) or actively supporting an injured astronaut. To work efficiently requires a certain level of autonomy of the robotic platform that can, in this way, be supervised by a human being for the high level decisions only. To share goals with the human requires also a deep confidence in the robotic platform safety as the robot must, in this situation, operate side by side with a human being.

Robotic platforms in support to human exploration are required to extend human capabilities and improve efficiency to reach the main goals of:

- constructing human habitats;
- monitoring and maintaining habitat environments;
- exploring the surface and perform scientific analysis;
- taking care of the astronauts' health.

In particular, a robot assistant like Eurobot would be involved in tasks such as:

- collect, transport, manipulate and assembly materials and components;
- move and inspect terrain, building maps;
- take photos and samples;
- store and analyze data.

As already suggested, potential utilization scenarios for Eurobot during a planetary surface mission include two separate phases, one relevant to "human arrival imminent" and the other concerning "human arrival".

3.2.1. Human Arrival Imminent

This scenario covers the typical activities to be performed by Eurobot on a planetary surface prior to the arrival of the human crew: most of the tasks are associated to the site exploration and preparation. Due to the constraints imposed by the time delays in the communications with the planet surface, interactions between the operators on Earth and Eurobot on the planet surface are limited and the robot should rely on a high autonomy in order to accomplish its goals. EGP required operations in such a scenario are mainly the following:

1. **Arrival and initial activation**. This phase covers the initial activities occurring right after surface landing, from Eurobot initial activation up to the time when the robot is driven off of the lander onto the planet surface. During this phase EGP systems are activated and checked and tests are performed to ensure that the rover is in a "safe state", before being moved for the first time beyond its original landing zone. Cameras are activated and antennas are deployed in order to allow the ground team to supervise Eurobot egress from the lander. Images of the surrounding terrain must be captured using EGP cameras; these images are transmitted in order to provide information regarding the egress status and to ensure that the surface directly beneath the rover's wheels does not present an immediate hazard: this allows the identification of a viable path from the lander. Rover egress, in fact, consists of a series of actions started by the ground only after previous step success verification: at this point of the mission terrain at the landing site is not well known, so the ground controller team must interactively assess EGP data and images, drive the rover, and then evaluate the resulting data before proceeding to subsequent steps. This iterative scenario needs to take into account the time delays between any command sent and the resulting feedback.

2. **Map generation and initial landing site exploration**. EGP cannot rely on a detailed and complete map of the surrounding environment (or DEM) as input before the start of the mission, as at the beginning it can have only knowledge about images or lidar measurements taken prior its landing. Thanks to the images acquired during outside operations, EGP is able to refine its map of the work site. After deployment and initial activation, images from the cameras are captured and processed into a set of (x, y, z) coordinates of points in front of the robot; the resulting map of points generates a set of terrain features (steps, slopes, roughness) which work as a detailed three-dimensional model of the actual terrain in front of the vehicle. This process is repeated while Eurobot moves along its path, thus building up a "world" model. With this map, EGP is able to determine if the terrain features represent obstacles for the rover and can as well select a potential safe path toward its destination, avoiding these obstacles. This feature is very useful to explore and precisely characterize the landing site area, both in preparation of the human arrival and to support a future landing risk assessment prior to the arrival of human supporting

infrastructures.

3. **Inspection of equipment**. Eurobot, through its cameras, can provide external views of the landed items, in order to confirm a nominal deployment and to allow the mission controllers to verify no damages occurred during the descent and landing phases. This may include the inspection of the external surfaces of the infrastructure and of specific items upon request from Earth. This applies both to the items landed together with the robot itself and to the following launches: the final descent and touch down phases are very critical and EGP, while on the surface, can provide additional and unique monitoring capabilities of the equipment. In addition, after its arrival on the planet surface, the robot can detect landed hardware accurate locations with respect to known landmarks and within the map that was generated.

4. **Landing site preparation and maintenance**. The landing site might require some preparation (which can be supported by Eurobot) like equipment deployment: many different types of tools are required to support human presence, such as antennas, cameras, spare parts, several instruments and so on. Eurobot can be used to deploy equipment from their stowage configuration or to relocate them wherever required. Mission controllers from Earth can order Eurobot to deploy/relocate a specific item to a target location within the site and Eurobot must be able to perform the required tasks to correctly prepare landing site. The grasping and transportation of an object can be executed autonomously by EGP: the item to be grasped can be identified through a dedicated target installed in a known position and then it can be stored on EGP payload tray, mounted on the rover, during transportation. This prototype can perform more complex tasks like manipulation of fluid, power, data lines and mating/de-mating of connectors or can also get access to areas requiring manipulation of MLI (opening and closure). The knowledge of the work site map allows it to move autonomously toward the target location, avoiding potential obstacles on its path. As a result of the inspections of the landing site, maintenance of some equipment may be required prior to the human arrival (e.g. exchange of failed equipment, replacement of MLIs, cleaning of surfaces from dust deposits) and can be supported by Eurobot.

3.2.2. Human Arrival

This scenario concerns the typical activities to be executed by the robotic assistant on a planetary surface after the arrival of the human crew: EGP has to cooperate with the humans to perform exploration tasks, to complete the construction of the site and to maintain it. During this phase, Eurobot has to go on performing the same type of activities with respect to pre-human arrival; in addition, the human presence on site leads to an increased interaction between the robot and the crew, thus allowing the execution of more complex tasks. Thanks to the human presence, in fact, cooperative operation such as the transportation of massive or big sized objects can be performed by Eurobot and EVA astronauts together; moreover, a direct control by an IVA operator with real-time monitoring (without any delay) is possible using an exoskeleton or a haptic device as interface. Therefore, in this phase, EGP main operations are:

1. **Support to EVA crew tasks**. Eurobot has to support the EVA crew during the execution of many activities on the planet surface. After the human arrival on the planet surface, the robot must be able to autonomously follow crew members to the target destination. It must keep a safe distance from the EVA crew walking on the surface, without slowing down the walk. EGP can respond to certain voice commands in order to execute specific actions (e.g. stopping, going to a definite location, acquiring images, taking samples, ...). This allows the EVA crew to interact with the robot without any additional support tool that might limit crew actions and that might be complex to be used with an EVA suit. In such a scenario, several activities require Eurobot support, for example:

    - transportation of equipment and tools: in spite of low gravity, astronauts might need EGP support for the transportation of some instruments that, at the same time, cannot be grasped by the robot itself. This is the case of massive or big sized objects, which require human-Eurobot cooperation for handling. The EVA members have the assignment to start this operation and then EGP can follow the crew steps while transporting the object. Eurobot in fact can react to pushing or pulling instructions given by the EVA crew member, while holding an object together with him;
    - grasping of samples: EVA crew members can require Eurobot to grasp a specific item or to take a geo-sample;

- exert a certain force and torque: in this way the robot can be used to multiply the crew member force and avoiding them to perform actions that may induce damages to their suits;
- EVA safety supporting functions: EGP can provide visual monitoring of the crew member, keep track of their sortie status (sortie time, resource status, location of the EVA crew within the map). It may monitor human health status and transfer a disabled EVA back to the base, if needed for safety reasons.

2. **Eurobot autonomous tasks**. The robot can be employed to autonomously inspect the human base, check its status and contribute to its maintenance over time through completion of actions such as:

   - discover and report (e.g. through proper images) any area of concern like a leakage or a visible damage;
   - keep and update an external item inventory, both in terms of visual status and location of all external tools and items;
   - inspect and evaluate the need for clean-up of dust sensitive areas just like airlock access;
   - execute specific maintenance tasks, such as exchange or relocation of equipment. In this case the operator should only indicate which equipment has to be grasped and Eurobot can autonomously perform all required operations;
   - exchange of EGP end-effectors;
   - manipulate material that is or has become hazardous for human beings;
   - reduce or even avoid crew intervention whenever external conditions are not suitable for men (e.g. during dust storms).

3. **Remote control from IVA operator**. For complex non routine activities, Eurobot can be "intuitively" controlled by a crew member inside a module on the planetary surface. This implies the use of an exoskeleton or a haptic device to control EGP arms; the operator can rely on the images from robot cameras in order to receive back useful feedback about his control actions. This kind of control may be required for repair activities (and in general for non standard procedures) which involve interfaces that are not fully robotically compatible.

3.3. EGP System Overview

The configuration of the ground prototype, a model of which is depicted in Figure 3.3.1, starts from the existing Eurobot concept for ISS application and its demonstrator, Eurobot Wet Model, and aims to adapt it to the planetary surface environment; this can be accomplished through the implementation of a wheeled rover for mobility. New functions on both autonomy and cooperation side are needed in order to demonstrate realistic tasks within test scenarios selected and designed to be representative of real exploration scenarios. Particular focus is also given to the interfaces between human and machine: for robotic assistant, in fact, interaction shall go beyond the traditional control station concept and experience more natural ways of control such as voice and gesture.

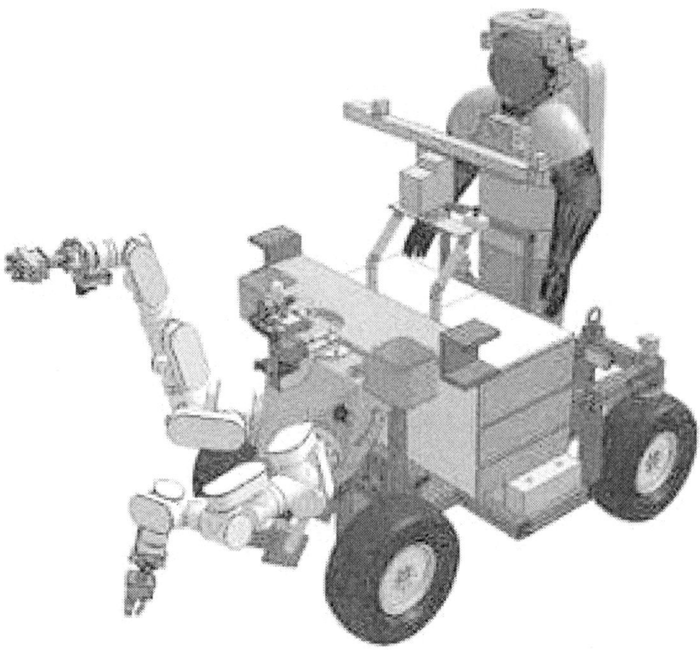

Figure 3.3.1. EGP model. Courtesy of ESA.

EGP is a mobile robot composed by a rover and two manipulator arms. Since the rover had to be extensively tested indoor in a limited area, it had to be able to steer around its main axis or in any case with small radius; in addition, it had to be able to run outdoor in "natural" terrain thus overcoming slight slopes and small steps. The speed of the rover is controllable up to a maximum of about $4-5\ km/h$ because EGP must follow a walking EVA without slowing down his walking. The size of the rover was defined accordingly to the requirement of accommodating a generic payload and a healthy EVA crew member, assumed standing and/or sitting. The rover also had to be able to transport a disabled EVA member, assumed laying and unconscious, but, in any case, EGP is not required to lift him on the rover. The final rover size was also affected by the necessary volume for avionic equipment allocation.

EGP two arms, which were directly taken from EWM, were accommodated in the top/front part of the rover; their position and orientation on the rover were selected to let the end-effectors reach and grasp an object on the ground, to allow the handover of items to and from an EVA astronaut standing nearby, to permit the robot to lift an equipment and to move it on the payload tray, either autonomously or cooperating with humans. Manipulation too has a great importance: EGP is able to manipulate robotics and non-robotics interfaces by the utilization of different self-exchangeable end-effectors that can be stored on a dedicated tray mounted on the rover. In particular, a two-fingered gripper was studied to be compatible with EVA interfaces, i.e. dog bone section handrails. Moreover, an electrical screwdriver for bolt screw operation was foreseen as second specialized end-effector, together with a three-fingered one for MLI handling and sample collection. These different end-effectors are shown in Figure 3.3.2.

Finer manipulation skills could be achieved with the integration of a "human" dexterous robot hand with separately controllable fingers. The autonomous grasping on the different objects/interfaces had to be supported by a proper vision system, which will be described in details in the following of this chapter.

Some of the tasks EGP must be able to execute consist, for example, in inspecting the landing site, autonomously reaching a designated location and transferring a disabled EVA back to the base with the shortest time or distance, moving in an environment where few information are known from the beginning (i.e. the Lander position and a rough map of the landing site).

Therefore, the prototype must build its own map and/or refine/modify its little a-priori knowledge; moreover, it must plan its path and be able to avoid collisions during the translation. EGP can detect human presence, using a proper target on the EVA suit, and follow an astronaut maintaining a safe distance,

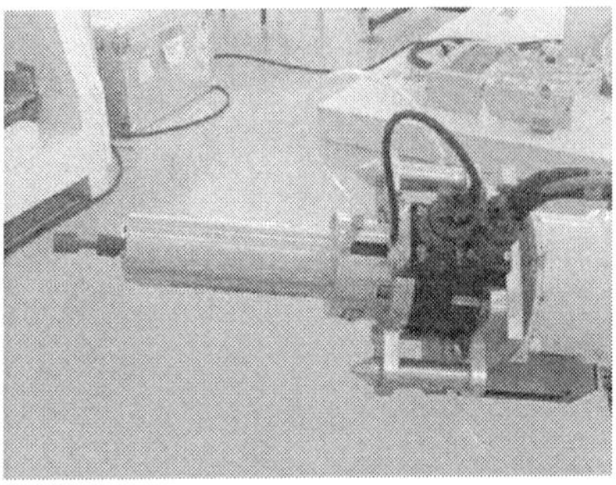

(a) electrical screwdriver.

(b) two-fingered gripper.

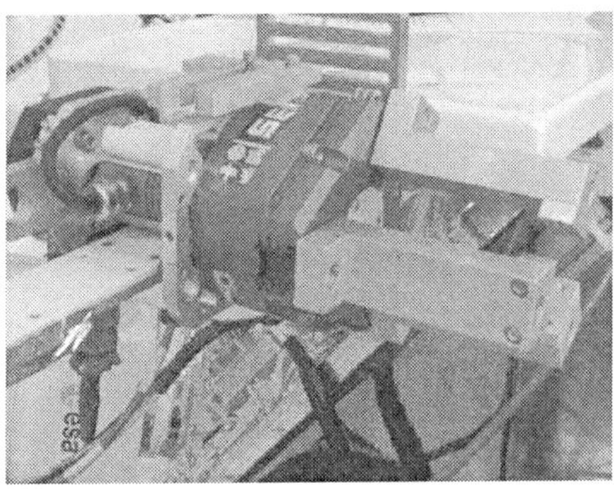

(c) three-fingered end-effector.

Figure 3.3.2. Different EGP end-effectors. Courtesy of ESA.

i.e. modulating its trajectory and speed on the basis of astronaut position and movements. Finally, the robot is able to physically cooperate with the astronaut, e.g. lifting an equipment together, following the movement of the human while providing strength.

Supervision control and manual mode are possible for safety reasons, both from a ground station and from "on surface"; for this last case, advanced MMI i.e. haptic device and exoskeleton for manipulation tasks requiring force feedback and voice control can be adopted. Figure 3.3.3 shows EGP in its demon-

Figure 3.3.3. EGP in its working environment during tests in Dutch Space. Courtesy of ESA.

stration facility at Dutch Space, in Leiden.

3.4. EGP Configuration

At the beginning of the study, an analysis was carried out to check the potential optimal utilization of the existing EWM hardware in EGP demonstrator, both from the wet and dry part of the equipment. The re-use of EWM hardware components was conceived in such a way to maintain the possibility to re-assemble the WET model in its original configuration and use it again in a neutral buoyancy facility. This means that single parts can be dismounted from EWM and utilized for EGP without an irreversible modification.

Two of the three EWM arms, fully integrated with all their electronics, were utilized for EGP: the whole EWM body structure has been accommodated on the rover with a 50° inclination; this orientation was selected in such a way to allow the robot manipulate objects that are on the ground. Beside range of motion, another considered criterion was linked to the total mass of the robot: the rover

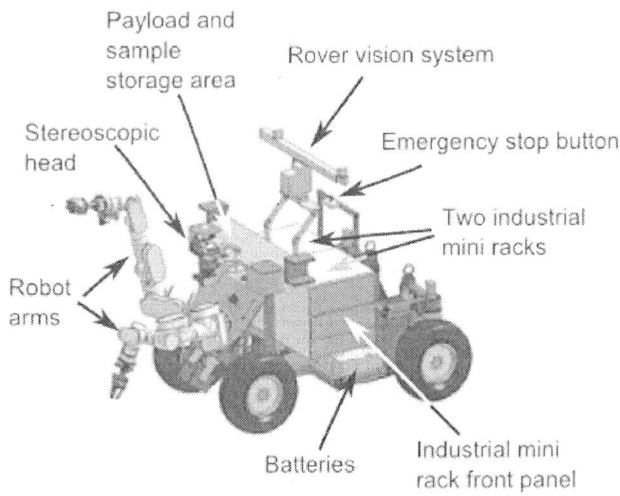

Figure 3.4.1. EGP main components. Courtesy of ESA.

must obviously keep its balance. Moreover, a suitable payload tray, mounted on its top part, allows EGP to store tools and samples, while carrying them. The rover is also endowed with an end-effector tool tray where the grippers not in use can be stored, in such a way to be located within EGP arm reach capability so to allow for an automatic end-effector exchange. Figure 3.4.1 shows some of the main Eurobot components.

The overall EWM vision system hardware, made up of the pan-tilt stereoscopic head and the wrist-mounted cameras, was re-used for EGP, in the same configuration: the stereo assembly was mounted on the top/front part of the rover, between the arms. Moreover, a new stereovision system was added to specifically support rover navigation, that is to acquire images from which the DEM can be built, and to provide images to human operators. Finally, the old dry segment part, that is the control station and the vision processing unit, were integrated onto the rover and powered by batteries, as the rover must be free to move.

The Robotic Control Station is the primary interface between the operator and EGP mobile segment: it forwards the operator issued commands to Eurobot GNC and receives back EGP telemetry from the EGNC; it also receives images coming from Eurobot arm cameras and from the rover stereo assembly. The rover and Eurobot manipulator arms have independent control layers: these last have priority over the vehicle layer. The rover navigation control system (RGNC) receives commands and telemetry only from the EGNC, which is able to ask the rover for help, in case it is too far from the object to be grasped. The EGNC, performing command and control of the arms, interfaces both with the VPU for image acquisition from the arm cameras and with the RGNC for command and telemetry routing. EGP platform mobility is thus controlled by the RGNC, on the basis of the external world data acquired by the pan/tilt stereo cameras. Figure 3.4.2 shows the realization of the overall robot, both rover and arms.

Figure 3.4.2. EGP at Thales Alenia Space, in Turin. Courtesy of ESA.

3.5. EGP Vision System

EGP overall vision system is made up of essentially two different subsystems: the rover stereo assembly for navigation control and the Eurobot stereo head and arm cameras for grasping and manipulation tasks. The main operations EGP must be able to execute and which can be accomplished thanks to its whole vision system are basically divided into two groups: some related to EGP arms and other to its mobile platform. Essential tasks consist in 3D object recognition, tracking and grasping, inspection, human recognition and tracking, 3D map environmental reconstruction and hazardous map building as support to navigation.

The manipulation stereo assembly (the one inherited from EWM and made up of three cameras) has the main job to allow human monitoring of EGP operations: images from these cameras are sent to the operator, who can easily supervise task execution. In such a way to make this monitoring possible, a particular control for the pan/tilt unit, accommodating this stereo assembly, was realized: the PTU is driven in such a way to follow, within its workspace limits, every end-effector movement, as highlighted by Figure 3.5.1.

In order to accomplish all these tasks, the vision system control segment is able to acquire images and control all camera parameters and status; whenever required by the EGNC, it can process acquired frames to search for known objects in the scene (e.g. ORU or other items for manipulation). The computed pose of the object of interest is sent back to the EGNC, as an answer to its request, and the robot can be controlled, through these data, to accomplish its grasping task. The vision system also supports inspection activities: still images and videos can be acquired by the arm cameras and sent to the human operator for evaluation, but EGP can also perform some autonomous inspection operations. On the other hand, the robot must be able to support autonomous navigation: through its pan/tilt stereo cameras it must generate a DEM, update its position on it and navigate in a safe mode, detecting hazards, avoiding obstacles and dangerous situations. Moreover, also human recognition and tracking is a distinctive feature of EGP vision system: this task, as the object grasping one, can be performed thanks to the presence of markers on the EVA suit. This solution is relatively simple to implement and allows the accomplishment of a very high reliability for the system; moreover, it turns out to be coherent with the future exploration scenarios.

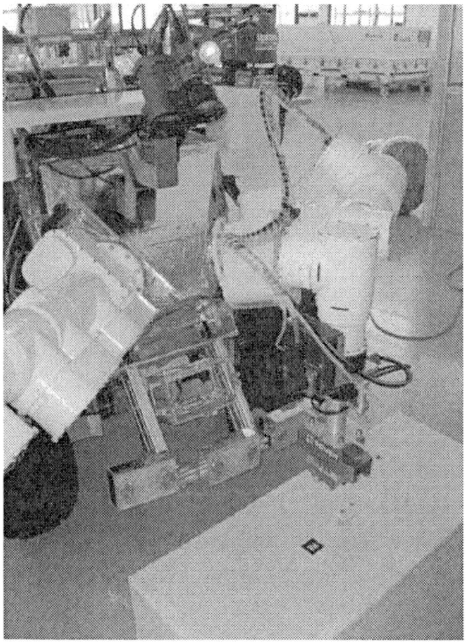

Figure 3.5.1. EGP manipulation task performed at Thales Alenia Space. Note that the stereo head follows end-effector motion, providing operation feedback. Courtesy of ESA.

Chapter 4

EGP Control Algorithms: Coordination from Force and Vision Feedback

4.1. General Control Architecture with Priority Tasks

In this chapter, a general strategy for coordination and control of complex robotic structures is presented. This technique originates from previous works conducted at Graal Lab, in Genoa, and fully illustrated in [34]. Indeed the control algorithm explained in the following has been developed within the context of the present Thesis and is a generalization of the work in [34]; it can be employed for complex robots, regardless of their own structure, as it will be clearer in a while.

This strategy, in fact, can be used for different situations: starting from a simple arm, until controlling an arm mounted on a moving base, as well as two (or even more) arms placed again onto a mobile platform. Furthermore, this technique can be applied to a general control case in order to achieve different control objectives: besides the accomplishment of the desired final velocity (or position) for the end-effector, any other control goals can be fulfilled. For example, requirements about robot safety can be determined: the arm posture can be controlled in order to maintain always a good manipulability or to avoid external obstacles (all these requirements can be assigned to different values of priority, according to the desired level of importance linked to each of them).

Each of these tasks starts with a desired velocity (that has been generated in different ways - for example through EGP vision system or force/torque sensors) and then the control action to accomplish this reference is computed, in accordance with the priority level of the currently considered task.

In the following, the simple general case of a fixed base arm is firstly presented and also its generalization to an arm mounted on a mobile platform is derived. Then a specialization of this coordination and control algorithm is presented, dealing with the specific EGP case. Finally, details on how reference velocities can be generated are given; in particular, EGP vision system and force feedback are analyzed.

4.1.1. Fixed Base Single-Arm with Priority Tasks

The first considered case concerns a single fixed base arm; dealing with the requirement to make the end-effector velocity vector $\dot{x} \triangleq \left[\omega^T, v^T\right]^T$ maximally similar to an assigned value $\bar{\dot{x}}$, note that this problem can be equivalently formulated as the following LQ optimization problem

$$\min_{\dot{q}} \|\bar{\dot{x}} - \dot{x}\|^2 \tag{4.1.1}$$

where $\dot{q}$ is the the joint velocity vector that induces a velocity $\dot{x}$ to the end-effector, according to the well-known relationship $\dot{x} = J\dot{q}$.

As already mentioned, besides the accomplishment of the desired final velocity, many other tasks can be required in order to avoid undesirable situations (obstacle hitting, manipulability losses,...). Hence, a sequence of desired tasks to be accomplished, as stated by Equation 4.1.2, can be listed:

$$S_1 \triangleq \left\{ \dot{q} = \underset{\dot{q}}{\operatorname{argmin}} \|\bar{\dot{x}}_1 - J_1\dot{q}\|^2 \right\} \tag{4.1.2a}$$

$$S_2 \triangleq \left\{ \dot{q} = \underset{\dot{q} \in S_1}{\operatorname{argmin}} \|\bar{\dot{x}}_2 - J_2\dot{q}\|^2 \right\} \tag{4.1.2b}$$

$$\vdots$$

$$S_i \triangleq \left\{ \dot{q} = \underset{\dot{q} \in S_{i-1}}{\operatorname{argmin}} \|\bar{\dot{x}}_i - J_i\dot{q}\|^2 \right\} \tag{4.1.2c}$$

$$\vdots \qquad i = 1, 2, \ldots, n$$

Each of these requirements is assigned to a different level of priority, decreasing while i increases. Starting from the first high priority task and defining

$$\dot{h}_0 = 0; \quad Q_0 = I \tag{4.1.3}$$

as an initialization, the control problem solution for the $i-th$ task is the following:

$$\begin{array}{ll} \bar{X}_i & = \bar{x}_i - J_i \dot{h}_{i-1} \\ \bar{J}_i & = J_i Q_{i-1} \\ \dot{\zeta}_i & = \bar{J}_i^{\#} \bar{X}_i \\ H_i & = \left(I - \bar{J}_i^{\#} \bar{J}_i\right) \\ \dot{h}_i & = \dot{h}_{i-1} + Q_{i-1} \dot{\zeta}_i \\ Q_i & = Q_{i-1} H_i \end{array} \tag{4.1.4}$$

Note that the expression of the desired $\dot{q}$ vector (for the $i-th$ task) is

$$\dot{q} = \dot{h}_i + Q_i \dot{z}_i \tag{4.1.5}$$

where $\dot{z}_i$ is a dummy vector and, if k is the least priority task, $\dot{z}_k = 0$. This means that, for $i = k$, no other control variables can be chosen in such a way to accomplish the desired motion: if the robot has not yet fulfilled all the requirements, the only possible solution consists in changing the reference commands.

Moreover, the error vector between desired and assumed velocity (that is the cost) reference can be computed as:

$$\|\dot{\varepsilon}_i^o\|^2 = \left\| \left(I - \bar{J}_i \bar{J}_i^{\#}\right) \bar{X}_i \right\|^2 \triangleq \left\| \bar{V}_i \bar{X}_i \right\|^2; \quad V_i \triangleq \left(I - \bar{J}_i \bar{J}_i^{\#}\right) \tag{4.1.6}$$

In the following a constructive proof is given. Considering the first task, it is possible to define the LQ optimization problem:

$$\|\dot{\varepsilon}_1^o\|^2 \triangleq \min_{\dot{q}} \|\bar{x}_1 - J_1 \dot{q}\|^2 \tag{4.1.7}$$

As the solution to Equation 4.1.7 can be trivially expressed as

$$\dot{q} = J_1^{\#} \bar{x}_1 \tag{4.1.8}$$

using Equation 4.1.8 the error relation can be rewritten as

$$\|\dot{\varepsilon}_1^o\|^2 = \left\| \left(I - J_1 J_1^{\#}\right) \bar{x}_1 \right\|^2 \triangleq \|V_1 \bar{x}_1\|^2 \tag{4.1.9}$$

where, obviously, the following definition has been made:

$$V_1 \triangleq \left(I - J_1 J_1^{\#}\right) \tag{4.1.10}$$

In such a way to take into account the not yet accomplished secondary tasks, a dummy vector $\dot{z}_1$ can be added; this vector must not affect the end-effector motion (otherwise it would change the final end-effector velocity and the control problem would not be fulfilled). Hence, a proper additional term has to be chosen, as in Equation 4.1.11: note that, in general, $\left(I - J_1^{\#} J_1\right)$ belongs to the null space $N(J)$ of J; this condition allows the second term of Equation 4.1.11 not to affect the primary task of goal frame reaching. For further details on these remarks see [34] and [35].

$$\begin{aligned} \dot{q} &= J_1^{\#}\bar{\dot{x}}_1 + \left(I - J_1^{\#} J_1\right)\dot{z}_1 = \\ &= J_1^{\#}\bar{\dot{x}}_1 + H_1 \dot{z}_1 \end{aligned} \tag{4.1.11}$$

Furthermore since, recalling definitions stated by Equation 4.1.3, the "equivalent velocity reference" $\bar{\dot{X}}_1$ can be written as in Equation 4.1.12

$$\bar{\dot{X}}_1 = \bar{\dot{x}}_1 - J_1 \dot{h}_0 = \bar{\dot{x}}_1 \tag{4.1.12}$$

and since

$$\bar{J}_1 = J_1 Q_0 = J_1 \tag{4.1.13}$$

the same problem can be formulated as in Equation 4.1.14:

$$\|\varepsilon_1^o\|^2 = \min_{\dot{q}} \left\| \bar{\dot{X}}_1 - \bar{J}_1 \dot{q} \right\|^2 \tag{4.1.14}$$

The expression for $\dot{q}$ retrieved from task 1 is thus

$$\begin{aligned} \dot{q} &= \bar{J}_1^{\#}\bar{\dot{X}}_1 + \left(I - \bar{J}_1^{\#}\bar{J}_1\right)\dot{z}_1 = \\ &= \dot{\zeta}_1 + H_1 \dot{z}_1 = \\ &= \dot{h}_0 + Q_0 \dot{\zeta}_1 + Q_0 H_1 \dot{z}_1 = \\ &= \dot{h}_1 + Q_1 \dot{z}_1 \,. \end{aligned} \tag{4.1.15}$$

where the last two lines formally hold since $\dot{h}_0$ and Q_0 have been initialized as in Equation 4.1.3. Obviously enough, the following assumptions have been made:

$$\begin{aligned} \dot{\zeta}_1 &= \bar{J}_1^{\#}\bar{\dot{X}}_1 \\ \dot{h}_1 &= \dot{h}_0 + Q_0 \dot{\zeta}_1 \\ Q_1 &= Q_0 H_1 \end{aligned} \tag{4.1.16}$$

Finally, the corresponding error vector for this primary task can be expressed as in Equation 4.1.17:

$$\|\dot{\varepsilon}_1^o\|^2 = \left\| \left(I - \bar{J}_1 \bar{J}_1^{\#}\right) \bar{X}_1 \right\|^2 \triangleq \left\| V_1 \bar{X}_1 \right\|^2 \tag{4.1.17}$$

Now, after having retrieved all these expressions, task 2 can be faced; for this last, the following LQ optimization problem has to be solved:

$$\|\dot{\varepsilon}_2^o\|^2 \triangleq \min_{\dot{q} \in S_1} \|\bar{x}_2 - J_2 \dot{q}\|^2 \tag{4.1.18}$$

From Equation 4.1.15 it is possible to explicitly expressed the requirement $\dot{q} \in S_1$ as:

$$\dot{q} \in S_1 \quad \Leftrightarrow \quad \dot{q} = \dot{h}_1 + Q_1 \dot{z}_1 \tag{4.1.19}$$

Hence, Equation 4.1.18 can be rewritten as

$$\begin{aligned} \|\dot{\varepsilon}_2^o\|^2 &= \min_{\dot{z}_1} \left\| \left(\bar{x}_2 - J_2 \dot{h}_1\right) - J_2 Q_1 \dot{z}_1 \right\|^2 = \\ &= \min_{\dot{z}_1} \left\| \bar{X}_2 - \bar{J}_2 \dot{z}_1 \right\|^2 \end{aligned} \tag{4.1.20}$$

where, similarly to Equations 4.1.12 and 4.1.13, the following expressions hold:

$$\bar{X}_2 = \bar{x}_2 - J_2 \dot{h}_1 \tag{4.1.21a}$$

$$\bar{J}_2 = J_2 Q_1 \tag{4.1.21b}$$

From Equation 4.1.20 the expression for $\dot{z}_1$ can be computed as

$$\begin{aligned} \dot{z}_1 &= \bar{J}_2^{\#} \bar{X}_2 + \left(I - \bar{J}_2^{\#} \bar{J}_2\right) \dot{z}_2 = \\ &= \dot{\zeta}_2 + H_2 \dot{z}_2 \end{aligned} \tag{4.1.22}$$

where definitions in Equation 4.1.23 has been made:

$$\begin{aligned} \dot{\zeta}_2 &= \bar{J}_2^{\#} \bar{X}_2 \\ H_2 &= \left(I - \bar{J}_2^{\#} \bar{J}_2\right) \end{aligned} \tag{4.1.23}$$

The expression for $\dot{q}$ can thus be retrieved as

$$\begin{aligned} \dot{q} &= \dot{h}_1 + Q_1 \left(\dot{\zeta}_2 + H_2 \dot{z}_2\right) = \\ &= \dot{h}_2 + Q_2 \dot{z}_2 \end{aligned} \tag{4.1.24}$$

where:

$$\begin{aligned}\dot{h}_2 &= \dot{h}_1 + Q_1\dot{\zeta}_2 \\ Q_2 &= Q_1 H_2\end{aligned} \quad (4.1.25)$$

Finally, using Equation 4.1.22, the cost is

$$\|\dot{\varepsilon}_2^o\|^2 = \left\|\left(I - \bar{J}_2\bar{J}_2^{\#}\right)\bar{X}_2\right\|^2 = \left\|\bar{V}_2\bar{X}_2\right\|^2 \quad (4.1.26)$$

where matrix $\bar{V}_2$ has been defined as:

$$\bar{V}_2 \triangleq \left(I - \bar{J}_2\bar{J}_2^{\#}\right) \quad (4.1.27)$$

Analogous maths can be made in order to generalize to successive tasks, thus retrieving, for the $i-th$ task, the formulation in Equations 4.1.4, 4.1.5 and 4.1.6.

4.1.2. Moving Base Single-Arm with Priority Tasks

In case the previous robotic arm is mounted on a supporting base, moving with a velocity $\dot{y}$ (both linear and angular), an approach based on Dynamic Programming technique can be assumed, as will be clearer soon. For this new robotic structure, the sequence of tasks shown by Equation 4.1.2 modifies as follows

$$\begin{aligned} &\vdots \\ A_i \triangleq &\left\{\dot{q} = \underset{\dot{q}\in A_{i-1}}{\operatorname{argmin}} \|(\bar{\dot{x}}_i - S_i\dot{y}) - J_i\dot{q}\|^2\right\} \\ &\vdots \qquad\qquad\qquad\qquad i = 1, 2, \ldots, n \end{aligned} \quad (4.1.28)$$

where S_i is the rigid body transformation, essential to compute the velocity induced to the the end-effector by the motion of the robot mobile base. Formally, simply defining

$$\bar{\dot{x}}_i^{\perp} \triangleq \bar{\dot{x}}_i - S_i\dot{y} \quad (4.1.29)$$

there is absolutely no problem to bring back this formulation to the previous one, with the same notation. The actual problem is, instead, the fact that the supporting base velocity $\dot{y}$ should appear explicitly as a parameter affecting both the cost and the control action. To achieve this aim, similarly to what done before, consider the primary task, for which the following problem can be stated:

$$\|\dot{\varepsilon}_1^o\|^2 \triangleq \min_{\dot{q}} \|(\bar{\dot{x}}_1 - S_1\dot{y}) - J_1\dot{q}\|^2 \quad (4.1.30)$$

The expression for $\dot{q}$ can be computed as in Equation 4.1.31, in which the additional dummy vector $\dot{z}_1$ has already been introduced (still not affecting the required primary task velocity accomplishment):

$$\begin{aligned}\dot{q} = &\ J_1^{\#}\bar{\dot{x}}_1 - J_1^{\#}S_1\dot{y} + \left(I - J_1^{\#}J_1\right)\dot{z}_1 = \\ &= J_1^{\#}\left(\bar{\dot{x}}_1 - S_1\dot{y}\right) + H_1\dot{z}_1\end{aligned} \tag{4.1.31}$$

Consequently, Equation 4.1.32 can be obtained:

$$\|\dot{\varepsilon}_1^o\|^2 = \left\|\left(I - J_1J_1^{\#}\right)\left(\bar{\dot{x}}_1 - S_1\dot{y}\right)\right\|^2 \triangleq \|V_1\left(\bar{\dot{x}}_1 - S_1\dot{y}\right)\|^2 \tag{4.1.32}$$

Similarly to the previous case, by simply initializing the quantities

$$\dot{h}_0 = 0; \quad Q_0 = I; \quad P_0 = 0 \tag{4.1.33}$$

the following notation can be formally introduced:

$$\begin{aligned}\bar{\dot{X}}_1 &= \bar{\dot{x}} - J_1\dot{h}_0 = \bar{\dot{x}}_1 \\ \bar{J}_1 &= J_1Q_0 = J_1 \\ \bar{S}_1 &= S_1 - J_1P_0\end{aligned} \tag{4.1.34}$$

Hence, the LQ optimization control problem modifies as

$$\|\dot{\varepsilon}_1^o\|^2 = \min_{\dot{q}}\left\|\left(\bar{\dot{X}}_1 - \bar{S}_1\dot{y}\right) - \bar{J}_1\dot{q}\right\|^2 \tag{4.1.35}$$

the expression for $\dot{q}$ is shown in Equation 4.1.36

$$\begin{aligned}\dot{q} &= \bar{J}_1^{\#}\bar{\dot{X}}_1 - \bar{J}_1^{\#}\bar{S}_1\dot{y} + \left(I - \bar{J}_1^{\#}\bar{J}_1\right)\dot{z}_1 = \\ &= \dot{\zeta}_1 - P_1\dot{y} + H_1\dot{z}_1 = \left(Q_0\dot{\zeta}_1 + \dot{h}_0\right) - P_1\dot{y} + Q_0H_1\dot{z}_1 = \\ &= \dot{h}_1 - P_1\dot{y} + Q_1\dot{z}_1 = \bar{J}_1^{\#}\left(\bar{\dot{X}}_1 - \bar{S}_1\dot{y}\right) + Q_1\dot{z}_1\end{aligned} \tag{4.1.36}$$

and the cost becomes:

$$\|\dot{\varepsilon}_1^o\|^2 = \left\|\left(I - \bar{J}_1\bar{J}_1^{\#}\right)\left(\bar{\dot{X}}_1 - \bar{S}_1\dot{y}\right)\right\|^2 = \left\|\bar{V}_1\left(\bar{\dot{X}}_1 - \bar{S}_1\dot{y}\right)\right\|^2 \tag{4.1.37}$$

The control problem concerning task 2 can be again stated as:

$$\|\dot{\varepsilon}_1^o\|^2 = \min_{\dot{q}\in A_1}\|\left(\bar{\dot{x}}_2 - S_2\dot{y}\right) - J_2\dot{q}\|^2 \tag{4.1.38}$$

Since, from the previous task

$$\dot{q} \in A_1 \quad \Leftrightarrow \quad \dot{q} = \dot{h}_1 - P_1\dot{y} + Q_1\dot{z}_1 \tag{4.1.39}$$

the problem modifies as follows:

$$\begin{aligned} \|\dot{\varepsilon}_1^o\|^2 &= \min_{\dot{z}_1} \left\| \left(\bar{\dot{x}}_2 - J_2\dot{h}_1\right) - \left(S_2 - J_2P_1\right)\dot{y} - J_2Q_1\dot{z}_1 \right\|^2 = \\ &= \min_{\dot{z}_1} \left\| \left(\bar{\dot{X}}_2 - \bar{S}_2\dot{y}\right) - \bar{J}_2\dot{z}_1 \right\|^2 \end{aligned} \tag{4.1.40}$$

Equation 4.1.40 leads to

$$\begin{aligned} \dot{z}_1 &= \bar{J}_2^{\#}\left(\bar{\dot{X}}_2 - \bar{S}_2\dot{y}\right) + \left(I - \bar{J}_2^{\#}\bar{J}_2\right)\dot{z}_2 = \\ &= \bar{J}_2^{\#}\bar{\dot{X}}_2 - \bar{J}_2^{\#}\bar{S}_2\dot{y} + \left(I - \bar{J}_2^{\#}\bar{J}_2\right)\dot{z}_2 = \\ &= \dot{\zeta}_2 - \bar{J}_2^{\#}\bar{S}_2\dot{y} + H_2\dot{z}_2 \end{aligned} \tag{4.1.41}$$

and, consequently, $\dot{q}$ can be written as:

$$\begin{aligned} \dot{q} &= \left(\dot{h}_1 + Q_1\dot{\zeta}_2\right) + \left(Q_1\bar{J}_2^{\#}\bar{S}_2 + P_1\right)\dot{y} + Q_1H_2\dot{z}_2 = \\ &= \dot{h}_2 - P_2\dot{y} + Q_2\dot{z}_2 \end{aligned} \tag{4.1.42}$$

Finally, the cost can be obtained as:

$$\|\dot{\varepsilon}_2^o\|^2 = \left\| \left(I - \bar{J}_2\bar{J}_2^{\#}\right)\left(\bar{\dot{X}}_2 - \bar{S}_2\dot{y}\right) \right\|^2 = \left\| \bar{V}_2\left(\bar{\dot{X}}_2 - \bar{S}_2\dot{y}\right) \right\|^2 \tag{4.1.43}$$

The same considerations can be made for all the other tasks, thus generalizing this approach; for the $i-th$ task the solution of the associated control problem is expressed by

$$\begin{aligned} \bar{\dot{X}}_i &= \bar{\dot{x}}_i - J_i\dot{h}_{i-1} \\ \bar{J}_i &= J_iQ_{i-1} \\ \dot{\zeta}_i &= \bar{J}_i^{\#}\bar{\dot{X}}_i \\ H_i &= \left(I - \bar{J}_i^{\#}\bar{J}_i\right) \\ \bar{S}_i &= S_i - J_iP_{i-1} \\ P_i &= P_{i-1} + Q_{i-1}\bar{J}_i^{\#}\bar{S}_i \\ \dot{h}_i &= \dot{h}_{i-1} + Q_{i-1}\dot{\zeta}_i \\ Q_i &= Q_{i-1}H_i \end{aligned} \tag{4.1.44}$$

and the expression for $\dot{q}$ and for the cost are given in Equations 4.1.45a and 4.1.45b respectively.

$$\dot{q} = \dot{h}_i - P_i\dot{y} + Q_i\dot{z}_i \tag{4.1.45a}$$

$$\|\dot{\varepsilon}_i^o\|^2 = \left\|\left(I - \overline{J_i J_i^{\#}}\right)\left(\overleftarrow{X}_i - \overline{S}_i\dot{y}\right)\right\|^2 = \tag{4.1.45b}$$

$$= \left\|\overline{V}_i\left(\overleftarrow{X}_i - \overline{S}_i\dot{y}\right)\right\|^2; \quad \overline{V}_i = \left(I - \overline{J_i J_i^{\#}}\right)$$

Note that no assumption has been made on the moving base velocity $\dot{y}$ (that is $\dot{y}$ is generally made up of both a linear and an angular part): this vector can be though of as the velocity of a generic robotic structure. For instance, in order to suggest how this technique can be applied to a specific structure like EGP, $\dot{y}$ can represent the velocity of the moving rover; however, this method is general enough to be simply extended to all other cases. Furthermore, the same approach can be easily extended to the coordination and control of more than two macro-structures (e.g. two arms mounted on a vehicle).

4.1.3. Final Remarks

As already mentioned, the presented control and coordination strategy can be extended to more complex cases in which multi-branched kinematic chains are involved; situations on both open (free arm motion) and constrained structures (end-effectors grasping an object) can be faced.

However, such a generalization (needed, for instance, to control and coordinate a dual-arm mobile manipulator) is not discussed in the context of this Thesis because planned space activities assigned a maximum priority level to the case of a single-arm mobile manipulator, already described.

4.2. EGP Coordination and Control: Single-Arm Mobile Manipulator Case

At this point, the presented control strategy is applied, in this specific case, to EGP. This last, as already said, is a dual-arm mobile manipulator endowed with a set of cameras (both mono and stereo) and force/torque sensors in such a way to be able to interact with the external environment. This complex robotic structure can be controlled in a coordinated manner, in order to accomplish hard tasks, requiring motion of both rover and arms. However, EGP main activities require that only one arm at a time is used and so the same approach just described is employed. As already said, if EGP main operations needed the coordination of both arms with the rover, the presented technique would remain valid and could be simply extended in such a way to control all the three macro-structures.

The control strategy proposed in this paragraph thus aims to obtain such a coordination between one arm and the vehicle, considering both of them as two separate macro-structures, each making its own contribution to motion. In such a way, it is possible to coordinate EGP arm and rover in order to both accomplish the required motion and also to add further secondary tasks, aiming to avoid bad situations happening during the task execution (e.g. assure a good manipulability during the arm motion).

With the purpose to coordinate vehicle and arm whenever EGP has to reach a desired final configuration or position, this proposed solution exploits *Dynamic Programming* potentialities [36], giving a two phase solution to the formulated problem. First, the robotic structure is considered from top to bottom (that is, first the arm and then the rover) and the control action for each macro-structure is retrieved, according to some criteria (*Backward Phase*). Then, by simply evaluating the found expressions, numerical values for each macro-structure motion contribution are computed (*Forward Phase*), thus controlling the overall system to the final desired position.

4.2.1. Backward Phase

First of all the Backward Phase (from top to bottom of the robot) is performed, in order to retrieve control actions for both arm and rover. The required motion can be subdivided into different tasks, adding also secondary *safety maintenance* requirements. Equation 4.2.1 shows a possible example of tasks for the coordination of the vehicle and of one arm, with this Dynamic Programming based strategy:

$$1)\ \bar{v} \approx (J_{L1}\dot{q}_1 + J_{L2}\dot{q}_2) \tag{4.2.1a}$$

$$2)\ \bar{\omega} \approx (J_{A1}\dot{q}_1 + J_{A2}\dot{q}_2) \tag{4.2.1b}$$

$$3)\ \bar{\mu} \approx p\dot{q}_1 \quad \leftarrow \quad \bar{\mu} \triangleq \lambda\mu \tag{4.2.1c}$$

The first two requirements are aimed to make the robot end-effector assume the desired cartesian reference velocity and accomplish required motion. Note that

Equation 4.2.1c represents a safety maintenance task: thanks to it, bad situations such as manipulability losses can be penalized and avoided.

The three tasks represented in Equation 4.2.1 have different priorities, in this case decreasing from the first until the last one. This is a fundamental aspect because it is possible to chose proper priority values according to the requirements that are to be more strictly satisfied. For instance, giving up the requirement about angular velocity or even requiring the fulfillment of a smaller subset of its components, it is possible to exploit extra degrees of freedom for secondary tasks (such as keeping a good robot posture or avoiding external object hitting). In fact, reducing these tasks to their essentiality turns out to be fundamental in order to improve the effectiveness of the other concurrent tasks. Examples of them are "always do the best in such a way to increase MM and keep it above an assigned threshold", or "always do the best in order to reduce the penalty functions assigned to the vicinity to any joint limit".

The term MM stands for Manipulability Measure and represents an index expressing the quality of a robot posture, evaluated according to some criteria. In fact, a situation in which, for example, the arm turns out to be almost completely stretched out must be strongly avoided as it could compromise the task completion. Hence, a function $\mu = f(\bullet)$ (in such a way to properly penalize situation to be avoided) is defined and represents the generic corrections introduced by safety maintenance tasks. Computing MM time derivative $\dot{\mu}$ the following relation is obtained

$$\dot{\mu} = \frac{d\mu}{dt} = p^T \dot{q} \tag{4.2.2}$$

where p is defined as:

$$p \triangleq \frac{\partial \mu}{\partial q} \tag{4.2.3}$$

From Equation 4.2.3 the direction k, along which the correction actions are exerted, can be computed as:

$$k = \frac{\partial \mu}{\partial x} = pJ^{\#} \tag{4.2.4}$$

Therefore, in general, resultant joint velocities can be written as

$$\bar{q} = J^{\#}\bar{x} + h(\bar{\mu} - k\bar{x}) \tag{4.2.5}$$

where h is defined as

$$h \triangleq \left[p\left(I - J^{\#}J\right)\right]^{\#} \tag{4.2.6}$$

and where the following relation holds:

$$\dot{\mu} = \bar{\mu} \triangleq \lambda\mu; \quad \lambda > 0 \tag{4.2.7}$$

Equation 4.2.7 fulfillment contributes to accomplish safety maintaining tasks, if the robot has a good initial posture. The so obtained expression for $\bar{q}$ is made up of two terms:

- $J^{\#}\bar{x}$, devoted to the primary task completion (i.e. execute required velocity reference commands);
- $h\left(\bar{\mu} - k\bar{x}\right)$, representing the secondary task contribution.

Note that, again, the vector h belongs to the null space $N(J)$ of J and so the second equation term does not affect the primary task of reaching the goal frame.

In order to solve the control problem and find the motion contributions of each macro-structure, starting from the desired Task 1) expressed by Equation 4.2.1a, the following relation can be written

$$\dot{q}_1 = J_{L1}^{\#}\left(\bar{v} - v_2\right) \tag{4.2.8}$$

in which v_2 (that is the rover finally accomplished velocity) plays the role of a parameter. In such a way to take into account further contributions to the other task achievement (and then also the rover action), it is possible to add to Equation 4.2.8 a further term, obtaining

$$\dot{q}_1 = J_{L1}^{\#}\left(\bar{v} - v_2\right) + \left(I - J_{L1}^{\#}J_{L1}\right)\dot{z}_1 \tag{4.2.9}$$

where $\dot{z}_1$ is an arbitrary vector. This term is added in such a way to better exploit available degrees of freedom to assure the fulfillment, as much as possible, of secondary tasks. Again, it does not affect the primary task, being $\left(I - J_{L1}^{\#}J_{L1}\right)$ an orthogonal projector onto the null space (or kernel) of J_{L1}. The vector $\dot{z}_1$ can thus be chosen in such a way to maintain a good manipulability, while the other term maximally approximates to the reference velocity vector. Through this step, $\dot{q}_1$ is obtained as a function of the linear velocity reference and is parametrized by the rover linear velocity and by the arbitrary $\dot{z}_1$ vector:

$$\dot{q}_1 = f\left(\bar{v}; v_2, \dot{z}_1\right) \tag{4.2.10}$$

After this computation, $\dot{q}_1$ found in Equation 4.2.9 must be replaced in Task 2) expression (Equation 4.2.1b) thus retrieving the following relation:

$$\bar{\omega} - \omega_2 = J_{A1} J_{L1}^{\#} (\bar{v} - v_2) + J_{A1} \left(I - J_{L1}^{\#} J_{L1} \right) \dot{z}_1 \tag{4.2.11}$$

Solving again the related least square problem, an expression for $\dot{z}_1$ can be written as:

$$\dot{z}_1 = f(\bar{v}, \bar{\omega}; v_2, \omega_2) \tag{4.2.12}$$

More in details, the entire relation that allows to compute $\dot{z}_1$ is shown in Equation 4.2.13:

$$\dot{z}_1 = \left[J_{A1} \left(I - J_{L1}^{\#} J_{L1} \right) \right]^{\#} \left[\bar{\omega} - \omega_2 - J_{A1} J_{L1}^{\#} (\bar{v} - v_2) \right] \tag{4.2.13}$$

Similarly to what done before, a further arbitrary term $\dot{z}_2$ can be introduced, thus obtaining the following expression for $\dot{z}_1$:

$$\begin{aligned} \dot{z}_1 &= \left[J_{A1} \left(I - J_{L1}^{\#} J_{L1} \right) \right]^{\#} \left[\bar{\omega} - \omega_2 - J_{A1} J_{L1}^{\#} (\bar{v} - v_2) \right] + \\ &+ \left\{ I - \left[J_{A1} \left(I - J_{L1}^{\#} J_{L1} \right) \right]^{\#} \left[J_{A1} \left(I - J_{L1}^{\#} J_{L1} \right) \right] \right\} \dot{z}_2 \end{aligned} \tag{4.2.14}$$

From Equation 4.2.14 it is clear that $\dot{z}_1$ depends also on the newly introduced vector $\dot{z}_2$ and can be written as:

$$\dot{z}_1 = f(\bar{v}, \bar{\omega}; v_2, \omega_2, \dot{z}_2) \tag{4.2.15}$$

Using Equation 4.2.14 in the expression for $\dot{q}_1$, the following relation can be retrieved, stating that the arm joint velocities, too, are function of $\dot{z}_2$:

$$\dot{q}_1 = f(\bar{v}, \bar{\omega}; v_2, \omega_2, \dot{z}_2) \tag{4.2.16}$$

After being able to express $\dot{q}_1$ as a function of variables and parameters belonging to the first two tasks, the third and last requirement can be exploited in order to compute the remaining $\dot{z}_2$. In fact, replacing $\dot{q}_1$ in Task 3) and using Equations 4.2.9 and 4.2.14, the following relation can be written

$$\begin{aligned} \bar{\mu} &= p J_{L1}^{\#} (\bar{v} - v_2) + \\ &+ p \left(I - J_{L1}^{\#} J_{L1} \right) H^{\#} \left[\bar{\omega} - \omega_2 - J_{A1} J_{L1}^{\#} (\bar{v} - v_2) \right] + \\ &+ p \left(I - J_{L1}^{\#} J_{L1} \right) \left(I - H^{\#} H \right) \dot{z}_2 \end{aligned} \tag{4.2.17}$$

where the H matrix has been defined as follows:

$$H \triangleq J_{A1} \left(I - J_{L1}^{\#} J_{L1} \right) \tag{4.2.18}$$

With the same computation as before and with the definition

$$K \triangleq p\left(I - J_{L1}^{\#} J_{L1}\right)\left(I - H^{\#} H\right) \tag{4.2.19}$$

the expression for $\dot{z}_2$ can be retrieved and expressed as:

$$\begin{aligned} \dot{z}_2 &= K^{\#}\bar{\mu} - K^{\#} p J_{L1}^{\#}\left(\bar{v} - v_2\right) + \\ &-K^{\#} p\left(I - J_{L1}^{\#} J_{L1}\right) H^{\#}\left[\bar{\omega} - \omega_2 - J_{A1} J_{L1}^{\#}\left(\bar{v} - v_2\right)\right] \end{aligned} \tag{4.2.20}$$

At this point, exploiting this relation among $\dot{z}_2$ and the reference vectors

$$\dot{z}_2 = f\left(\bar{v}, \bar{\omega}, \bar{\mu}; v_2, \omega_2\right) \tag{4.2.21}$$

and replacing the so computed expression for $\dot{z}_2$ in the equation defining $\dot{q}_1$, the following relation comes out for the arm joint velocities:

$$\dot{q}_1 = L_1\left(\bar{v}, \bar{\omega}, \bar{\mu}; v_2, \omega_2\right) \tag{4.2.22}$$

More in details, the expression for $\dot{q}_1$ is given by Equation 4.2.23

$$\begin{aligned} \dot{q}_1 &= J_{L1}^{\#}\left(\bar{v} - v_2\right) + AH^{\#}\left[\bar{\omega} - \omega_2 - J_{A1} J_{L1}^{\#}\left(\bar{v} - v_2\right)\right] + \\ &+ABK^{\#}\bar{\mu} - ABK^{\#} p J_{L1}^{\#}\left(\bar{v} - v_2\right) + \\ &-ABK^{\#} pAH^{\#}\left[\bar{\omega} - \omega_2 - J_{A1} J_{L1}^{\#}\left(\bar{v} - v_2\right)\right] \end{aligned} \tag{4.2.23}$$

where the following definitions were assumed:

$$\begin{aligned} A &= \left(I - J_{L1}^{\#} J_{L1}\right) \\ B &= \left(I - H^{\#} H\right) \end{aligned} \tag{4.2.24}$$

Thanks to these first steps of the Backward Phase, it has been possible to compute an expression for a suitable control action (for the arm), which maximally satisfies the three required tasks. At this point, the rover contribution to the structure motion is still unknown; replacing the expression in Equation 4.2.22 in the initial tasks defined by Equation 4.2.1, the following relationships can be obtained:

$$1)\ \bar{v} \approx J_{L1} L_1\left(\bar{v}, \bar{\omega}, \bar{\mu}; v_2, \omega_2\right) + v_2 \tag{4.2.25a}$$

$$2)\ \bar{\omega} \approx J_{A1} L_1\left(\bar{v}, \bar{\omega}, \bar{\mu}; v_2, \omega_2\right) + \omega_2 \tag{4.2.25b}$$

$$3)\ \bar{\mu} \approx p L_1\left(\bar{v}, \bar{\omega}, \bar{\mu}; v_2, \omega_2\right) \tag{4.2.25c}$$

Moreover, as v_2 and ω_2 are linked to $\dot{q}_2$ through the following

$$\begin{aligned} v_2 &= J_{L2}\dot{q}_2 \\ \omega_2 &= J_{A2}\dot{q}_2 \end{aligned} \tag{4.2.26}$$

it is possible to retrieve the initial expressions of the successive iteration:

$$1)\ \bar{v} \approx \tilde{J}_{L2}\left(\bar{v}, \bar{\omega}, \bar{\mu}; \dot{q}_2\right) \tag{4.2.27a}$$

$$2)\ \bar{\omega} \approx \tilde{J}_{A2}\left(\bar{v}, \bar{\omega}, \bar{\mu}; \dot{q}_2\right) \tag{4.2.27b}$$

$$3)\ \bar{\mu} \approx \tilde{p}\left(\bar{v}, \bar{\omega}, \bar{\mu}; \dot{q}_2\right) \tag{4.2.27c}$$

This is the starting point for the second part of the Backward Phase: applying again the same procedure used in order to find the correct arm motion (contribution that maximize the achievement of required tasks), the rover velocity $\dot{q}_2$ can be similarly computed, thus obtaining:

$$\dot{q}_2 = L_2\left(\bar{v}, \bar{\omega}, \bar{\mu}\right) \tag{4.2.28}$$

During this procedure, the still unused degrees of freedom are exploited and a convenient rover motion is chosen, in order to maximally satisfy the potentially not yet accomplished tasks. Within this computation, a MM-monitoring turns out to be essential to understand if a manipulability loss situation happens, so that the control loop could behave accordingly to this event, as will be clear in a while.

4.2.2. Forward Phase

In this phase, evaluation of the found expressions is performed, thus finally retrieving the control actions that allow to reach the desired position. In fact, knowing the reference commands $\bar{v}$, $\bar{\omega}$ and $\bar{\mu}$, $\dot{q}_2$ can be correctly computed and, from it, v_2 and ω_2 too, thanks to Equation 4.2.26. At this point, by simply applying these relations to the equation obtained for $\dot{q}_1$, the arm joint velocity can be expressed in terms of $\bar{v}$, $\bar{\omega}$ and $\bar{\mu}$ only, and so evaluated.

Note that, even if this procedure finds the best possible $\dot{q}_1$ and $\dot{q}_2$ with respect to the required task fulfillment, the risk of eventually having MM losses still exists, depending on the required end-effector motion. This happens, for instance, when the object must be very high lifted: in such a case the platform cannot compensate the motion. Hence, MM must be constantly checked: if at the end of the Backward Phase the MM-monitoring highlights an increasing approaching of MM toward μ_0, only one action is possible. In fact, as there is no other macro-structure that can help the task achievement thanks to its motion, the obvious last-resort solution is modulating the linear velocity reference $\bar{v}$ (and, potentially, $\bar{\omega}$ too) in such a way to invert this tendency. This implicitly means a "priority change" implementation.

The overall Dynamic Programming approach is based on this mechanism, as the control loop cannot allow motion causing losses of manipulability: in order to prevent such an occasion, the control law must give up reaching the original goal and change the reference commands. This automatic priority change is accomplished through the following procedure: during the Backward Phase, besides $\dot{q}_2$ computation, $\dot{\mu}$, too, is obtained as:

$$\dot{\mu} = L_3\left(\bar{v}, \bar{\omega}, \bar{\mu}\right) \tag{4.2.29}$$

Hence, as $\dot{\mu}$ must be maximally similar to $\bar{\mu}$, in order to fulfill Equation 4.2.30

$$\bar{\mu} \approx L_3\left(\bar{v}, \bar{\omega}, \bar{\mu}\right) \tag{4.2.30}$$

the old reference command is modified by the introduction of a new velocity vector $\hat{\bar{v}}$; the so obtained reference vector $\tilde{v}$ (the last-resort solution) is expressed by Equation 4.2.31:

$$\tilde{v} = (1-\alpha)\,\bar{v} + \alpha\hat{\bar{v}} \tag{4.2.31}$$

Just like before, from this reference vector it is possible to compute $\dot{q}_2$ as

$$\dot{q}_2 = L_2\left(\tilde{v}, \bar{\omega}, \bar{\mu}\right) \tag{4.2.32}$$

and then go on with the following Forward Phase.

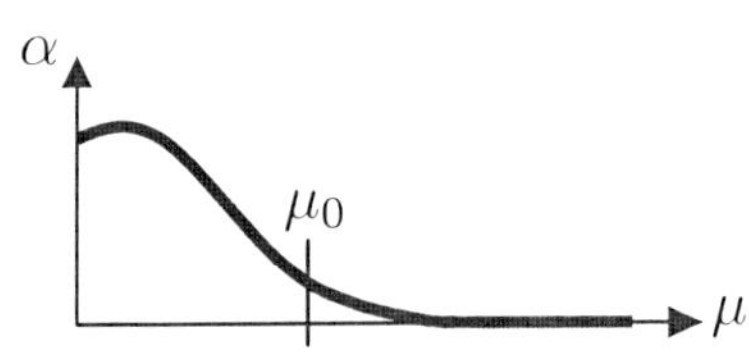

Figure 4.2.1. Alpha gain curve trend.

The α gain is responsible for the priority change: when $\alpha = 0$ the robot has a good manipulability and $\tilde{v} \equiv \bar{v}$, where as if $\alpha = 1$ MM is too low and $\tilde{v} \equiv \hat{\bar{v}}$, which is the correction velocity term to increase MM. Figure 4.2.1 shows the trend of this gain.

This approach is surely effective, as, besides allowing to accomplish required motion, it can prevent bad situations; moreover, it can be applied to any kind of robot: if there are more macro-structures than only two to be coordinated, the described procedure can be simply generalized to fit the specific robotic system. Besides, other secondary tasks can be added: for instance, as already suggested, the structure can be further asked to avoid hitting against external obstacles.

4.3. Vision-Based Recognition of Objects

Thanks to the developed general control architecture, it is possible to effectively coordinate EGP arms and rover. The Dynamic Programming based technique, as already illustrated, is able to start from a desired velocity and to coordinate the robot until the achievement of its tasks. In this sense, with the purpose to completely solve the addressed control problem, one issue is still open: the velocity reference generation. One of the possible method to compute such a required reference is based on EGP vision system; in fact, EGP is able to recognize objects of interest for its mission goals and locate them in the 3D space. After the object position has been retrieved, by simply comparing it with the desired final position the robot must assume with respect to the object itself, a velocity reference is obtained; giving this vector as requirement to the control algorithm previously detailed, it is possible to correctly achieve the desired coordination and control for the robot. In this paragraph the vision system is described and the overall image processing chain is highly detailed.

EGP scenarios plan that the arms have to interact with several objects present in the scene. The system is provided with an internal database containing the nominal position of each object (handles, launch fittings, MLI and so on). However this information is not sufficient to perform the required task, since the expected errors from the arm and rover positioning can prevent the task completion. For this reason a vision-based algorithm was set up in order to refine the object approaching phase: its objective is that of getting the position of the interesting item, in a 3D environment, thus allowing EGP to precisely identify its position. The recognition of general 3D objects is not so easy and robust: there are problems such as the need of a precise geometrical model for each object, the possible difficulty to choose the right item if several similar objects are present in the scene, the deformations that can occur because of different points of view and that can make the object seem different (thus leading to a failure in the recognition task) and so on. In order to avoid such problems, allowing EGP to be able to interact with the environment through vision, a method based on fiducial markers of the type shown in Figure 4.3.1 was used: each object of interest is associated to a different marker, univocally recognizable by the system, and its position in 3D space (6 degrees of freedom, 3 for linear position and 3 for orientation) is computed in such a way to control the robot to grasp the marked object. Each marker is a square planar pattern with known dimensions and structure and is assigned to a unique identification number: this allows the robot to recognize and locate more than one object. For this purpose an open-source software library, named *ARToolKitPlus*, was selected; it provides some functionalities for marker pose estimation and is robust enough, in terms of precision and velocity with which the position is computed, in order to be used in a real-time control loop. Moreover, another open-source library, named *OpenCV*, was exploited for all image processing functions [3].

Figure 4.3.1. Example of ARToolKitPlus fiducial marker. Image credits: Graz University of Technology.

ARToolKitPlus is a *C* and *C++* language software library developed for Augmented Reality applications, namely, dealing with applications that have to overlap virtual computer graphic images on the real world. One of the most difficult parts of developing an Augmented Reality application is precisely calculating the user's viewpoint in real time; ARToolKitPlus uses computer vision techniques to calculate the real camera position and orientation relative to marked cards. As a collection of software libraries, designed to be linked into application programs, it must be compiled on the specific operating system and platform. In order for the library to be used, it is also necessary to have some markers available (some default markers are provided by the library itself) and a minimum frame rate of 15 fps is recommended (below that frame rate the performance of ARToolKitPlus rendering module would be too restricted). Moreover, best performance is achieved if each marker is somehow kept flat. For further details on ARToolKitPlus see [4], [5], [6] and [7].

In order for the marker to be tracked, its entire black square border must be in view and also the pattern contained within the square border must be visible at all times; in such a situation, if there are troubles in the pose estimation procedure, it may be because of lighting conditions. This particular issue can often be fixed by changing the lighting threshold value used by the image processing routines.

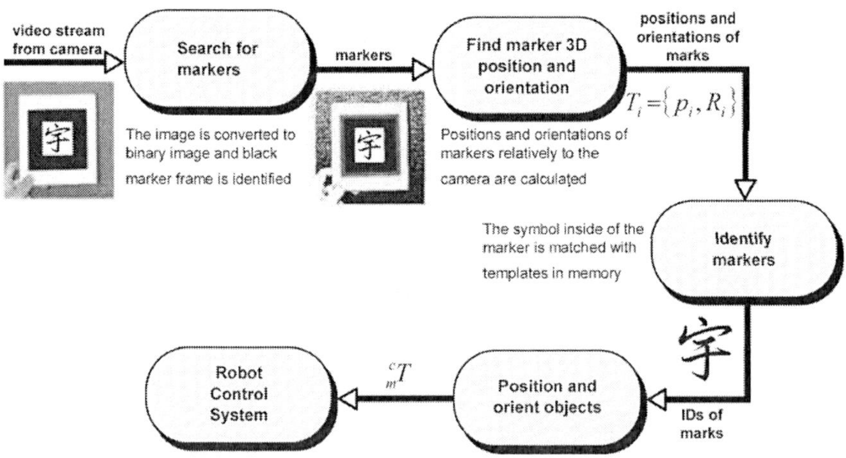

Figure 4.3.2. ARToolKitPlus procedure steps. Image credits: Graz University of Technology.

The ARToolKitPlus-based tracking system works as follows:

- the camera captures a video stream of the real world making it available for the system;
- the robot software searches through each video frame for any square shapes;
- if a square is found, the software uses some mathematics to calculate the position of the camera relative to the black square;
- the pattern inside is then matched with templates stored in memory and the marker ID is found;
- the computed position for that specific object is then used to generate a reference for the robot.

Figure 4.3.2 summarizes these steps.

For the image processing to be successful, some coordinate transformations are necessary, in order to move from the distorted screen coordinates and to end up knowing the relation between the marker and the camera; these transformations are displayed by Figure 4.3.3.

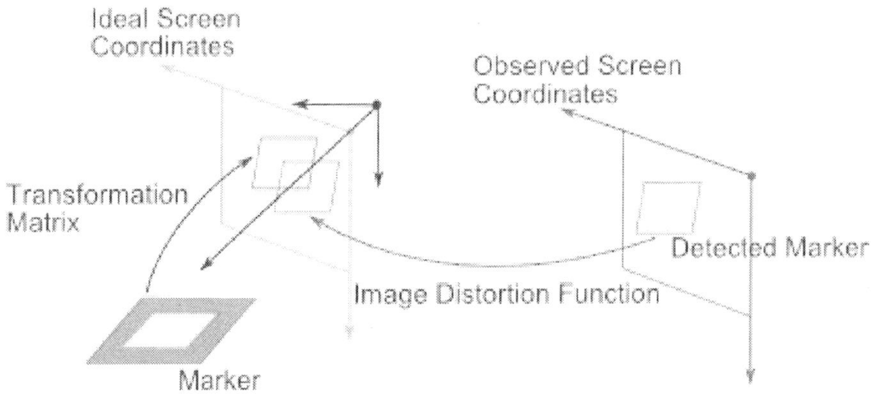

Figure 4.3.3. ARToolKitPlus main coordinate systems. Image credits: Graz University of Technology.

There are some limitations to this tracking: as already said, it is possible only if markers are in view and this also means that if a part of the pattern is somehow hidden the tracking will fail. There are also range issues: the larger the physical marker, the further away the pattern can be detected. Table 4.3.1 shows some typical maximum ranges for square markers of different sizes.

Table 4.3.1. Tracking range for different sized patterns. Credits: Graz University of Technology

Pattern size (cm)	Usable Range (cm)
7	40.6
8.9	63.5
10.8	86.4
18.7	127

This range is also affected somehow by pattern complexity: the simpler the marker, the better; patterns with large black and white regions (i.e. low frequency ones) are the most effective. Marker orientation in front of the camera is also an important issue for successful pattern identification: as the marker grows more tilted and horizontal with respect to the camera, the pattern becomes less visible, thus making recognition more unreliable. Finally, tracking results are also affected by lighting conditions: overhead lights may create reflections and glare spots on a paper marker, thus making the search for its borders very hard; to reduce this effect more non-reflective materials should be used. Figure 4.3.4 shows some graphs that underline the error variation with respect to different measured parameters.

4.3.1. Tracking Algorithm and Robot Control

The basic tracking algorithm structure is very simple: the video streaming from the camera is available for the application that searches for one or more known markers into each frame. When at least one marker is found, the procedure goes on computing the transformation matrix between the camera and the marker itself. This information is used by the robot control loop to drive the arm end-effector in a desired position, controlling also its orientation. Whenever the required motion demands an intervention from the rover, the previously described algorithm can be exploited to coordinate the overall system. This procedure is summarized by Table 4.3.2.

Table 4.3.2. Tracking algorithm steps

Initialization	1. Initialize the video capture and read in the camera parameters.
Main Loop	2. Grab a video input frame.
	3. Detect markers and recognize patterns in the grabbed video input frame.
	4. Compute the camera transformation relative to the detected patterns.
	5. Send the pose estimation to the robot control loop.
Shutdown	6. Close the video capture down.

Steps 2 through 5 are repeated continuously until the application quits, while steps 1 and 6 are just performed on initialization and shutdown of the application, respectively. Figure 4.3.5 shows in details the image processing that is

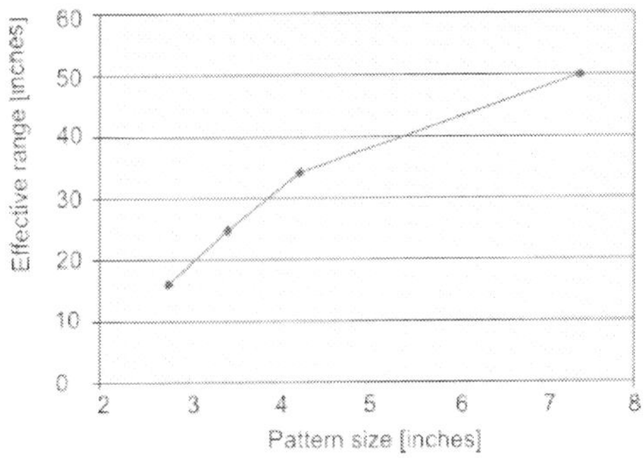

(a) effective range with respect to pattern size.

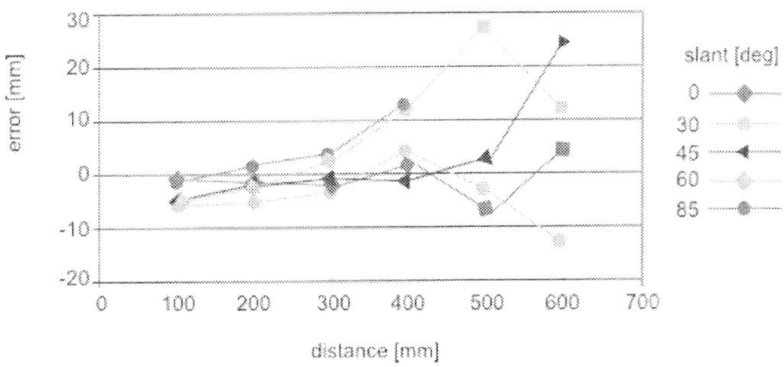

(b) error with respect to the distance from the camera.

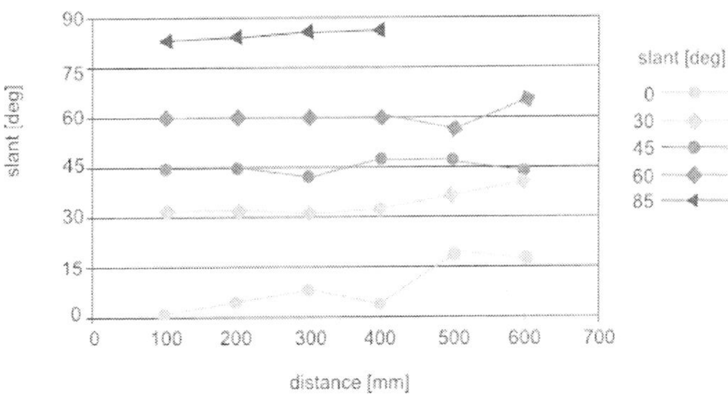

(c) tracking error with respect to angle.

Figure 4.3.4. Tracking error and range with respect to different parameters. Credits: Graz University of Technology.

needed in order to get the marker to be tracked.

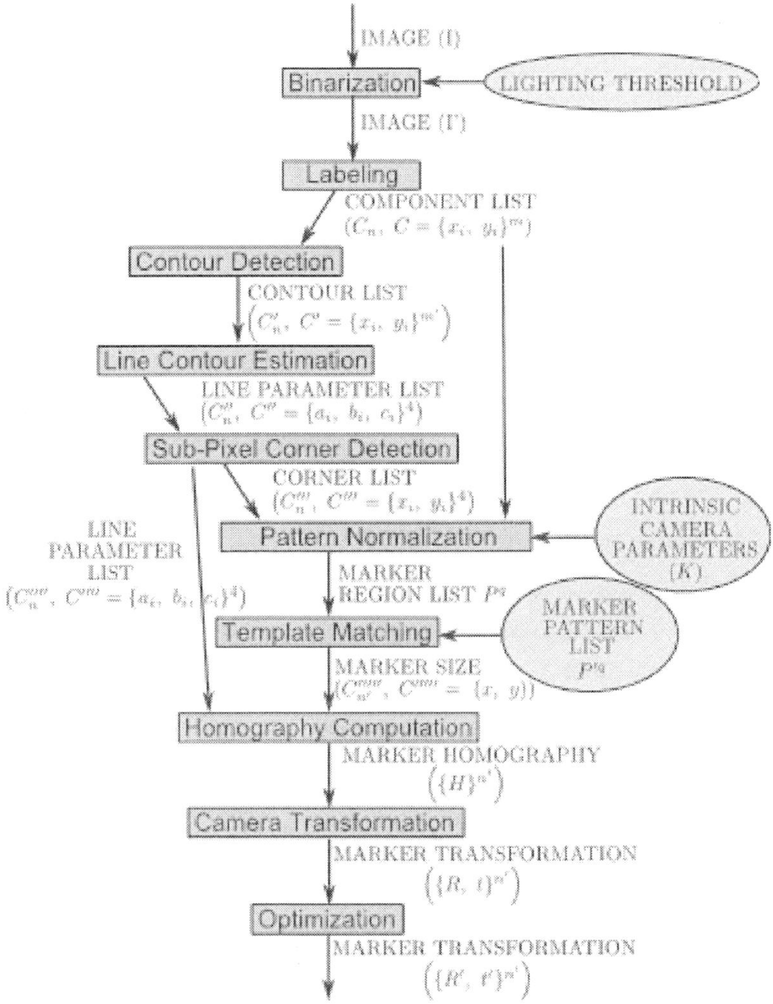

Figure 4.3.5. Details of the tracking procedure. Image credits: Graz University of Technology.

More in details, the input image is first binarized and processed in order to retrieve the marker contours; this information is then used to compute the equations of the lines delimiting the pattern (each one is represented by its coefficients a, b, c, such that its equation is expressed as $ax + by + c = 0$). The next step consists in corner detection with sub-pixel precision, in order to get the four marker corners. Then the pattern is normalized, using the known camera parameters, and a template matching is executed, among all patterns stored in the system memory. Finally, after some computations (including also an optimization procedure), the transformation matrix between camera and marker is retrieved. In order to increase the tracking stability, the library provides a so called history function that enables the estimator to use information from the previous image frame (in order to reduce marker jittering) instead of using only the current one to compute the position of the marked object. While using this history function, the result is less accurate because timing performances are increased at the expense of accuracy.

This is, in general, how a simple marker pose estimator works; however, considering the high level of autonomy and reliability desired for Eurobot, some more algorithms were analyzed and included in the vision control software, in order to face hostile conditions preventing the estimator from accomplishing its computation.

The tracking procedure, in fact, is highly affected by issues such as:

- lighting conditions: too bad illumination can affect estimation quality as it may create reflections and glare spots, thus worsening the pose estimation procedure results;
- complexity of the scene: the system may get confused by too many different objects in the surrounding environment (maybe some elements, after being binarized in the image, come to be similar to known patterns...);
- distance from which the marker is seen: this is linked to the time needed to retrieve the estimation; the more distant is the marker, the longer time is needed.

These limitations can be overcome, adding some appropriate steps (summarized in Table 4.3.3) to the tracking algorithm such as an accurate threshold procedure and an image cleaning and zooming. In fact, when the system receives the command to search for a marker (with the specification of which arm Ű and so which camera Ű is to be used and which marker is to be found), it acquires the corresponding arm mounted camera and in each frame it starts searching for the desired pattern. If the estimator finds one or more candidate markers in the frame, it does not compute their positions yet, as this processing is the most computationally heavy phase. So, the system proceeds to clean the image up: only the pixels corresponding to candidate markers are kept and a temporary output image is built up, equal to the original one in the area of possible candidate markers and white elsewhere.

Table 4.3.3. Tracking algorithm with thresholding, cleaning and zooming steps

Initialization	1. Initialize the video capture and read in the camera parameters.
Main Loop	2. Grab a video input frame.
	3. Apply some different threshold values to the image and search for a candidate marker. If the marker has been found, apply the same threshold to all following frames.
	4. Clean the image up and leave only areas containing candidate markers. If the estimation error is below a threshold, zoom the image and hand in it to the estimator.
	5. Detect markers and recognize patterns in the image. Compute the camera transformation relative to the detected patterns.
	6. Apply a low-pass filter to the estimation error.
	7. Send the pose estimation to the robot control loop
Shutdown	8. Close the video capture down.

This improves the recognition in the sense that the complexity of the scene is highly reduced after this processing. Then, the actual estimation is done (using the cleaned image as input of the estimator) and both linear position and orientation of the object to be grasped are obtained. The arm is then controlled in such a way to stop right in front of the marker and to keep a predetermined distance from it.

During this marker centering phase, when the arm has almost reached the desired position, if the error (both in terms of linear and angular components) between the estimated and the desired pose falls below a specific threshold, the image, after being cleaned up, is zoomed in its center: in this way the estimation computation time decreases. This aspect is important because the achievement of a faster dynamics for the estimation means that the variations in the estimated pose come to be at higher frequency and can then be removed with the aid of a simple low-pass filter; in this way annoying oscillations of the arm while accomplishing its motion can be prevented. Figure 4.3.6 highlights the difference of dynamics between the estimation error in the zoomed and non-zoomed cases.

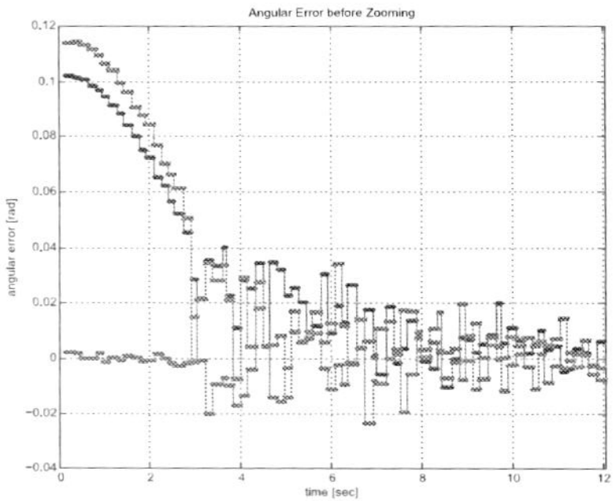

(a) angular error before zooming.

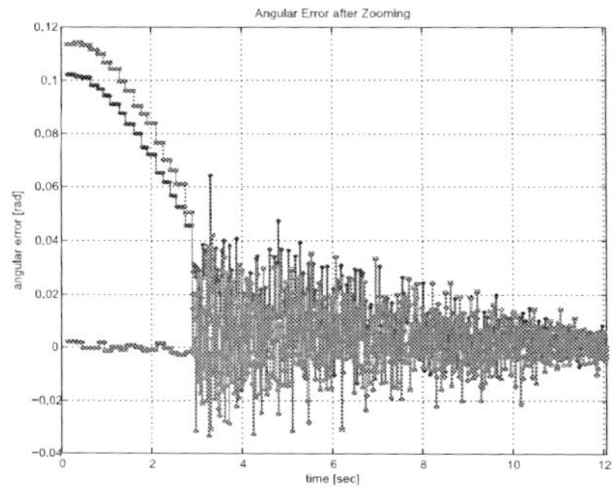

(b) angular error after zooming.

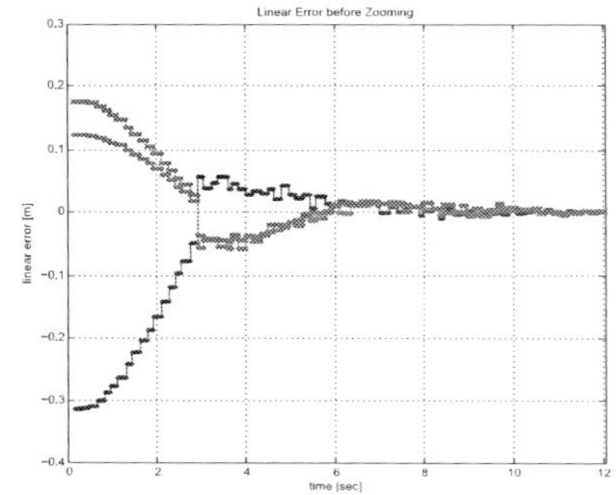

(c) linear error before zooming.

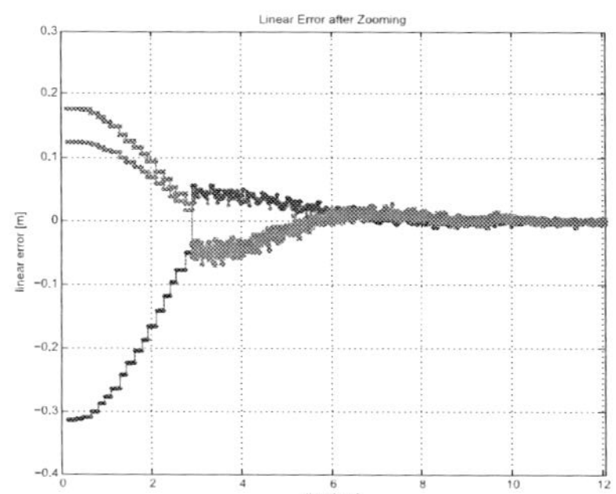

(d) linear error after zooming.

Figure 4.3.6. Estimation error dynamics.

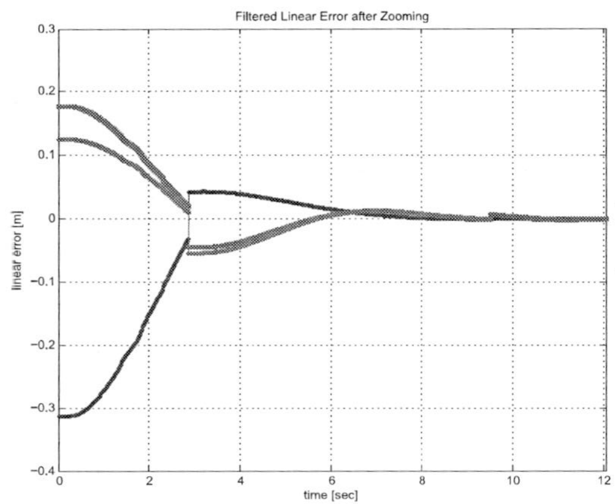

(a) linear error.

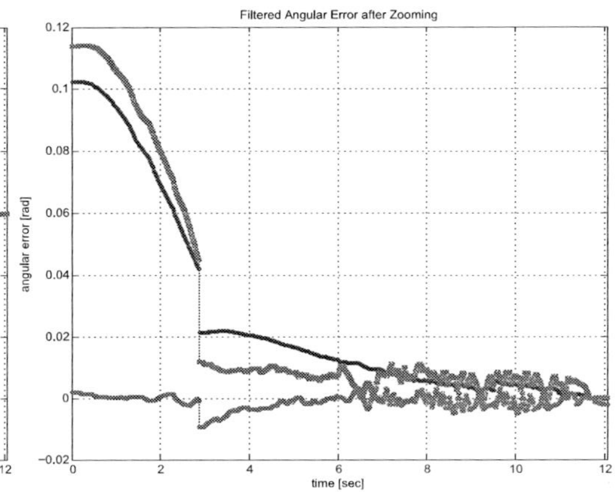

(b) angular error.

Figure 4.3.7. Filtered estimation error.

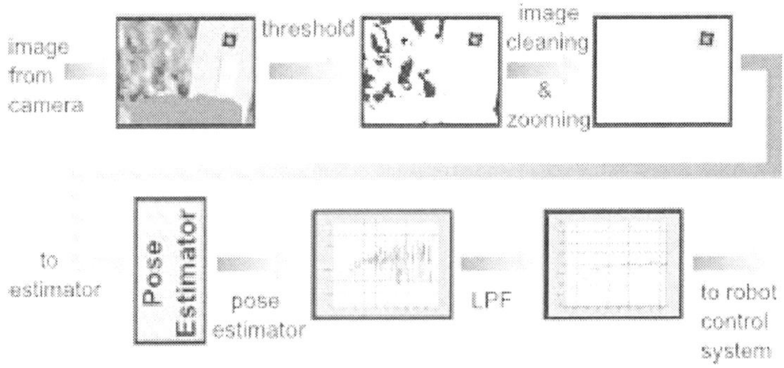

Figure 4.3.8. Vision-based tracking steps.

Obviously, while zooming the image, the estimated linear components are smaller than the actual ones, as to the system the marker appears to be closer; so also the reference linear position must be changed: for example, using a zooming factor of $2\times$, the linear component of the reference must be halved.

Figure 4.3.7 shows the action of a low-pass filter, if applied to the estimation error. Finally, as previously mentioned, the estimator is also sensitive to the lighting conditions: with a too bright illumination but also with a low light in the scene, the algorithm may fail the marker recognition. In order to avoid such a situation, a preliminary thresholding is executed: before all other computations, the estimator receives the input image, converts it to black and white and applies to it a threshold value for luminance, in order to increase contrast. After this procedure, if the marker is found, the same threshold is applied to all the other frames and the tracking algorithm goes on with the previously mentioned steps; otherwise, other threshold values, among some previously defined ones, are tried until the marker is recognized. The whole search for marker is summarized in Figure 4.3.8, in which all the necessary steps and their outputs can be seen.

While setting up the vision algorithm, several attempts have been made before ending up with the previously described final procedure. In particular, a color-based image processing was tested, in order to further improve the already defined vision strategy: this was done in order to prevent the binarization step and the external conditions from confounding the system during pattern recognition. To remove this chance, thus improving tracking stability with respect to bad illumination, the idea is simple: instead of black and white markers, colored ones can be employed (e.g. green and white). Then a new algorithm was implemented: input image was converted to HSV color space and then a series of thresholds was applied.

For each image pixel, the H component was thresholded using two simple upper and lower bounds; only the pixels with a H value in the correct range were kept in the output image. This was done in order to select the right color (in this case green) of the space. The S and V components were filtered in a quite different way: for each pixel the quantity expressed in Equation 4.3.1 was computed and compared with a threshold, named Th_c; the corresponding pixel was kept in the output image only if the c value was below the specified threshold.

$$c = \sqrt{S^2 + V^2} < Th_c \tag{4.3.1}$$

This step removed from the image all black and white pixels, together with all the too dark or too light green ones. Once found the green regions to be kept, through well-known blob analysis algorithms, the original image was masked with the only areas of interest, thus getting an image equal to the original one in the surroundings of the green regions, and white elsewhere. Starting from this new image, marker estimation was performed as before.

This solution was conceived in order to assure an improvement in the camera white balancing and in the marker detection itself, because of the compensation for not perfect lighting conditions thus made: eliminating too dark and too light pixels helped to obtain a more uniform luminance in the image. Table 4.3.4 shows the overall algorithm steps, including this color-based procedure.

Table 4.3.4. Tracking algorithm steps, including green detection

Initialization	1. Initialize the video capture and read in the camera parameters.
Main Loop	2. Grab a video input frame.
	3. Convert image to HSV colour space. Consider only the green pixels and change the rest to white
	4. If the estimation error is below a threshold, zoom the image and hand in it to the estimator.
	5. Detect markers and recognize patterns in the image. Compute the camera transformation relative to the detected patterns.
	6. Apply a low-pass filter to the estimation error.
	7. Send the pose estimation to the robot control loop
Shutdown	8. Close the video capture down.

Nevertheless, this color-based algorithm was excluded from the final vision system image processing, mainly for the following reasons:

- the stability tracking improvement is not so remarkable;
- thresholds are very hard to be properly tuned: perceived color hue changed too much according to illumination;
- the computational time required for this color-based procedure is too much: this is because the algorithm had to process each image pixel by pixel in order to discard non-green areas.

While searching for a known pattern in the scene, the system computes the marker pose with respect to the camera ${}^{c}_{m}T$ (and so also the target object location is known, as the relative positioning between each marker and its assigned object is known and constant); the desired position ${}^{c^*}_{m}T$ that the camera (and so the end-effector) must assume with respect to the marker is a priori-known: for example the robot is required to stop right in front of the marker, keeping a previously defined distance from it.

From the computed transformation matrix ${}^{c}_{m}T$ it is possible to build up matrix ${}^{w}_{c^*}T$, representing the desired final position: considering the relative positioning of the main EGP frames, depicted in Figure 4.3.9, it can be expressed as

$$ {}^{w}_{c^*}T = {}^{w}_{c}T {}^{c}_{m}T \left({}^{c^*}_{m}T \right)^{-1} \tag{4.3.2} $$

where ${}^{c^*}_{m}T$ is known and ${}^{w}_{c}T$ can be obtained from ${}^{e}_{c}T$ and ${}^{w}_{e}T$: the first one is known and constant, since it only depends on where the camera is mounted on the arm wrist, while the other is computed at each sample time by the robot control system. Once obtained this reference matrix ${}^{w}_{c^*}T$, as said before, it is possible to retrieve a reference velocity vector and send it as an input to the Dynamic Programming based control strategy.

4.4. Force-Based Approaching Toward Objects

Besides the vision system, also the JR3 force/torque sensors can help in the reference velocity generation. Within the context of a grasping task, for instance, when the robot has to approach the object and actually grasp it, a further velocity can be required to the end-effector, according to the detected contact between

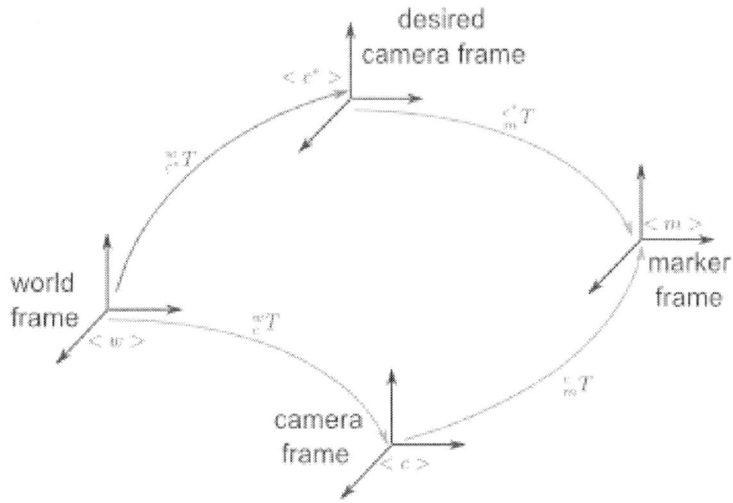

Figure 4.3.9. Relative positioning of EGP main frames.

gripper and object. Then, again, the previously Dynamic Programming based algorithm can be exploited.

As soon as the object has been recognized and located in the 3D space and the alignment between the robot arm and the object itself is accomplished, the successive approaching toward this item and its actual grasping are still to be performed. To this aim, a direct force control strategy is used (for details refer to [1] and [2]), in order to zero potential residual errors caused by the vision system. Once the marker has been centered, the arm begins to approach the object and, when a contact is detected, the force control starts its action, generating a desired velocity. The hypothesis that only a pure force is applied to the contact point has been made; as this contact point does not coincide with the sensor location, the JR3, mounted on the arm wrist, will sense both a force f_s and a torque τ_s. These components are compared to desired force and torque references and from the so computed error the velocity to be assigned to the end-effector is obtained. In order to get this velocity to correctly drive the arm end-effector, the estimation of the contact point comes to be an essential step; this point, represented by the vector r, appears within the relation among torque, force and lever arm vector

$$\tau_s = [-f_s \wedge] r \tag{4.4.1}$$

where $[-f_s \wedge]$ is the skew-symmetric matrix corresponding to the vector $-f_s$ and

r is the unknown. This equation is not sufficient in order to estimate the contact point, as the determinant of the matrix $[-f_s\wedge]$ is null and so Equation 4.4.1 has infinite solutions; another condition may hence be added in order to solve the system: the only acceptable solution belongs to the palm surface, as the gripper must touch the object. So the contact point can only have a known distance d from the sensor and can thus be expressed as

$$r = [r_x \; r_y \; d] \tag{4.4.2}$$

where r_x and r_y are the unknown x and y components of the contact point and the z axis of the reference frame is along the arm, positive outgoing from the end-effector.

So, the contact point can be estimated solving Equations 4.4.1 and 4.4.2 together and, once computed, is used as the new tool center point in order to control the robot: the velocity $\bar{x}_c$ to be assigned to the contact point can be expressed as

$$\bar{x}_c = K\left(F_s^* - F_s\right) \tag{4.4.3}$$

where F_s^* and F_s are the generalized vectors of torque and force reference and feedback, respectively, while the gain matrix K is a diagonal 6×6 matrix with different tunable values for angular and linear velocities.

All these algorithms (both for what concerns force and vision based approaches and for the coordination and control technique) combine to bring about the final correct EGP behavior in both autonomous and cooperative tasks. The next chapter reports all the results achieved from both simulative and experimental tests conducted on EGP.

Chapter 5

EGP Simulative and Experimental Results

5.1. Autonomy and Cooperation

Among all the operations Eurobot must be able to execute, two main directions can be identified: *autonomy* and *cooperation*. The most important tasks for such a robotic demonstrator, in fact, are linked to the capabilities to both replace astronauts in some activities and help them during difficult operations. As a proof of the effectiveness of such an autonomous and cooperative robot, mainly two tasks were selected as representative of real exploration scenarios. Firstly, a grasping task stating the efficiency of EGP autonomy was demonstrated; then, also the cooperative lifting of equipment together with astronauts was considered highly representative since its completion assures an essential support to the human crew. A further operation, slightly different from the other manipulation tasks, was chosen to demonstrate again EGP autonomy: the solar array deployment. All these operations were thus selected in order to be accomplished by EGP demonstrator and will be described in the following.

5.1.1. EGP Autonomous Grasping Task

Within its grasping task, EGP must simply locate a specific object (an ORU or a MLI or some other items) and autonomously manipulate it. In such a way to make the robot interact with the surrounding environment, the vision algorithm previously described is exploited.

This manipulation skill is fundamental because it allows to accomplish more complex autonomous operations, such as the replacement of broken equipment, consisting of many different grasping sub-actions.

Simulative Results

In order to test the vision system, a preliminary simple interface was realized, using OpenGL functionalities; in this way the algorithm has been validated and tested under ideal conditions for the camera (no distortion due to its lenses, perfect illumination, absence of noise due to real images). This interface shows the surrounding environment from the robot arm camera point of view and has been used as a proof of concept in earlier algorithm evaluation. Simulations (from which the image sequence in Figure 5.1.1 is taken) show the grasping of a simple ORU: if its corresponding marker is found in the scene, the control system drives the end-effector in such a way to make the camera assume the desired position. When it has been reached, a new phase begins and consists in making the robot gripper approach the object of interest, in this case the ORU handle, and grasp it.

As already mentioned, the interface simulates a video capture as made by the robot itself (with the aid of a camera mounted onto its end-effector) while searching for a marker in the scene. The robot is controlled through the visual feedback until it assumes a position from which it could see the marker aligned to the red square drawn in figure. This square represents the desired view that the camera must reach and is computed through the knowledge of parameters such as the camera focal length, the marker size and the desired distance to keep between camera and marker. Then, after the marker has been centered by the robot and the camera is positioned as requested, a new phase begins, namely the approaching phase, and makes the robot end-effector approach the yellow handle and grasp it.

This interface does not simulate the whole tracking algorithm, but it only estimates the marker pose, without all the procedures of cleaning and zooming previously described. So, in order to prove the overall tracking phase, a more complex simulator was realized; it has two different views: one showing the overall robot (rover and arms) and the other displaying what is seen by the end-effector mounted camera. This simulator has a reference control window from which the robot can be controlled in different ways, for example sending either joint or cartesian references, or choosing the robot tool center point (e.g. camera

(a) Initial position.

(b) Centering phase and correction in the orientation.

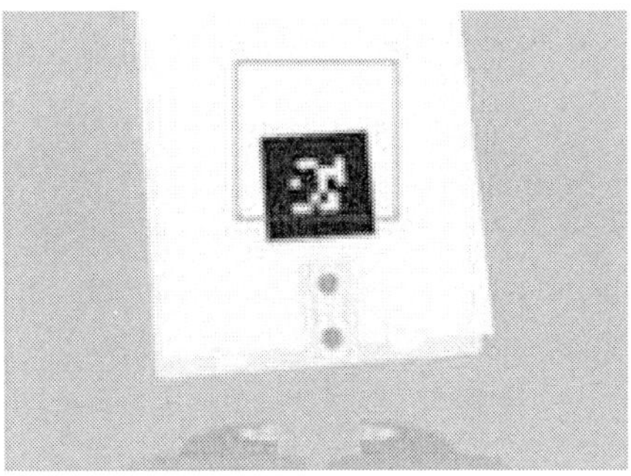

(c) Centering phase and correction in the orientation.

(d) Centered marker.

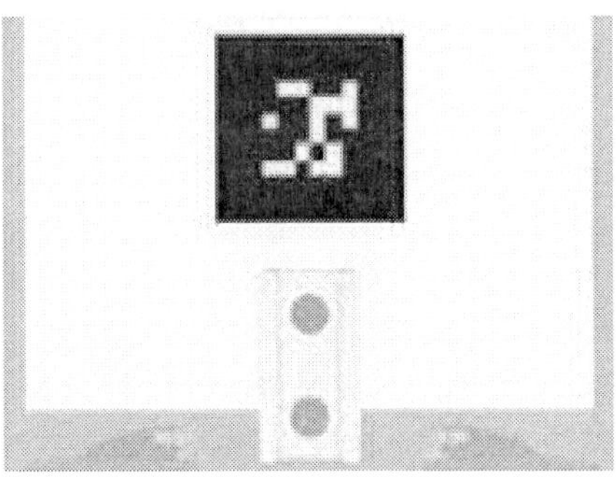

(e) Approaching toward the handle.

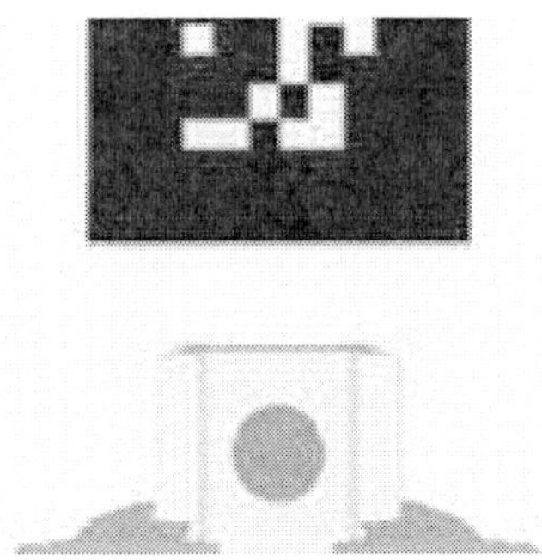

(f) Grasped handle.

Figure 5.1.1. Vision algorithm earlier evaluation. Courtesy of ESA.

or end-effector), or determining which task has to be simulated. Figure 5.1.2

shows both the reference control console and the view of the overall robot; note that this last provides a 3D visualization of EGP and can thus arranged in such a way to observe the task simulation from different points of view.

Thanks to it, the entire task can be simulated within a more realistic scenario: the complete final tracking algorithm is entirely implemented and evaluated; the robot is thus controlled through visual feedback. Moreover, this simulator takes into account the planetary scenario and tests the correct grasping of an ORU placed on the planet surface.

Figure 5.1.3 shows the accomplishment of the described grasping task in the simulations on the overall vision algorithm, realized thanks to this second more complex simulator.

Experimental Results

Many tests were performed, both at Thales Alenia Space in Turin and at Dutch Space in Leiden (The Netherlands). The first tests, in Turin, were aimed to evaluate the performance of the vision system only, in order to understand if the image processing was robust enough to face real conditions such as bad illumination, noisy camera frames, complex scene and so on. The effectiveness of this vision algorithm was confirmed by these preliminary experimental results, in terms of both convergence velocity and precision. In fact the robot was able to center the marker within few seconds, thus assuring a fast task execution; moreover, a good precision is achieved as the vision guided phase ends up with a maximum linear error smaller than 0.15 cm and an angular error below 0.03 radians per second (i.e. more or less 1.72 degrees).

As these tests gave positive results, the implementation went on with the evaluation of the force control strategy: as already illustrated, a simple direct force control loop was realized, in such a way to try to accomplish the actual grasping phase. Thanks to the precision obtained from the previous vision based control strategy, the force phase can begin with small residual errors to be compensated. This allows the robot to correctly grasp the ORU even if some limitations have been highlighted. The major problem is due to EGP end-effector length: on each arm, the JR3 force/torque sensor is mounted on the wrist and the contact with external objects occurs at the palm level. As the distance between wrist and palm is remarkable, the force feedback cannot be very accurate and, in some cases, it can be even confused with the sensor noise.

For the test campaign, in order to prove Eurobot effectiveness within an

(a) 3D EGP visualization.

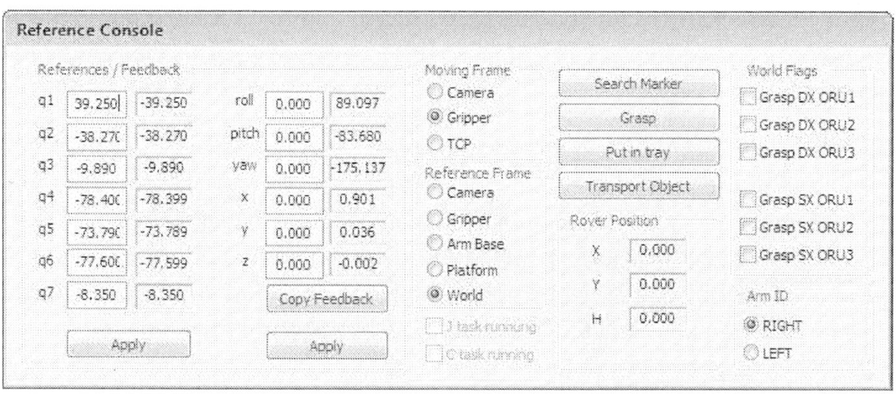

(b) Reference command window.

Figure 5.1.2. Reference console and 3D robot view of the realized complex simulator. Courtesy of ESA.

(a) initial position with errors.

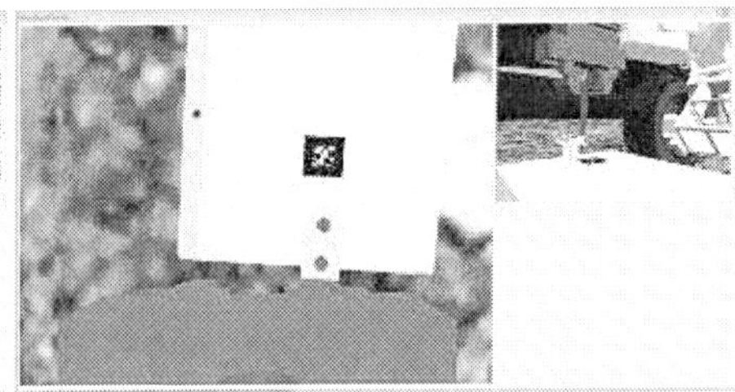

(b) marker centering.

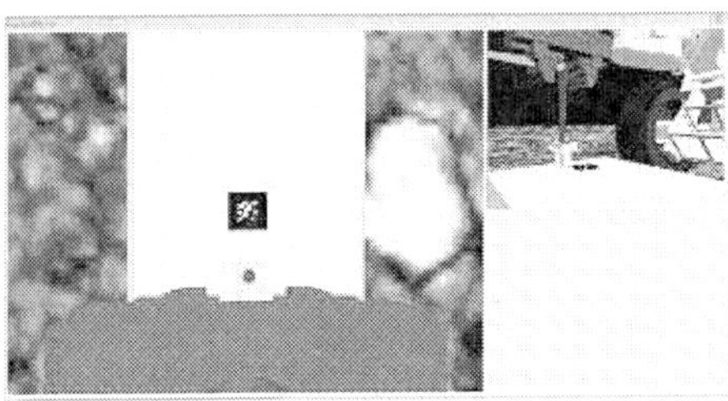

(c) last position corrections.

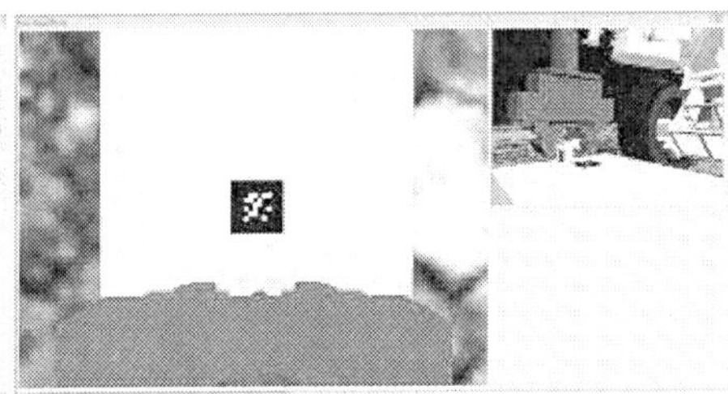

(d) grasped handle

Figure 5.1.3. Two view simulator. Courtesy of ESA.

exploration scenario as much real as possible, an EGP Test Playground Support Equipment was conceived. It includes all the items required to support a valid demonstration such as

- **Lander simulator**, including several robotic interfaces: it is supposed be "just landed" and requires localization, inspection and operations by EGP. It includes various items simulating cargo and equipment, and various robotic interfaces in support of manipulation task demonstration. A model of the lander simulator is shown in Figure 5.1.4, while the actual realized one is depicted in Figure 5.1.5; it houses in the following items:
  - removable cargo/equipment boxes;
  - MLI cover;
  - a deployable cable with connector;
  - a deployable structure (solar array);
  - a panel with connectors;

- an access panel fixed by screws;
- launch fittings;
- an item/panel with visible damage.

Equipment cargo are to be removed and transported by EGP; a deployable cable is included to allow the demonstration of complex tasks like the deploy of lines/cables and connector mating/de-mating. After deployment, the connector will be plugged into an antenna support base, previously transported by EGP. The dummy deployable structure in Figure 5.1.4 represents a solar array; for this item, the required deployment task is such that EGP rover too shall move to complete the action (this operation will be detailed in the following). Finally, a damaged panel is located in the higher part of the Lander, in order to verify the performance of the EGP cameras during the inspection tasks.

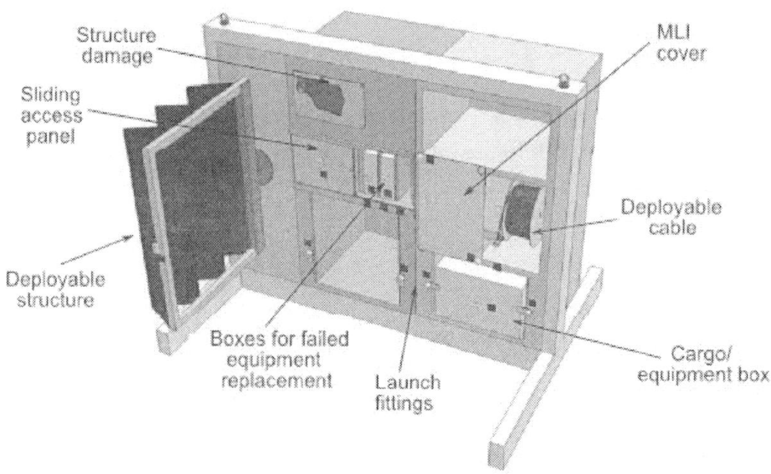

Figure 5.1.4. Model of EGP Lander simulator provided with dummy equipment. Courtesy of ESA.

- **Dummy equipment boxes**: boxes of different dimensions were realized. The smallest ones were designed for removal/replacement manipulation tasks within the Lander; the medium boxes were planned to be used for removal from the lander and transportation to selected locations; the larger

Figure 5.1.5. Lander simulator realized for EGP demonstration. Dummy obstacles, belonging to the ETPSE, simulate rock presence in EGP working background and white equipment boxes can be identified onto the Lander. Courtesy of ESA.

ones were conceived for cooperative tasks i.e. lifting the box from the ground together with the astronaut. For handling and transportation each box is provided with a dog-bone shaped interface (used by the robot) and with EVA handrails, attached to the sides of larger boxes (for human grasping).

- **Dummy antenna**: an antenna is a possible example of an item that needs to be unloaded, transported and maybe assembled with robotic support. The dummy antenna is composed of three parts: a base box with dummy connectors, a parabolic disk and a support column.
- **Dummy obstacles**: they are required in order to simulate rocks and to try out the autonomous navigation system.

Preliminary tests were conducted in Turin and highlighted EGP capability

Figure 5.1.6. ORU grasping task tested in Turin. Courtesy of ESA

to grasp a simple ORU placed on the ground, near the Lander; frames in Figure 5.1.6 show the execution of such a task.

Right after the execution of the first preliminary test at Thales Alenia Space, EGP was led to Dutch Space, in Leiden, for the final tests and demonstration. There a working environment was set up, in order to simulate a planetary scenario for EGP tests. Figure 5.1.7 shows such a working environment.

Thanks to the completion of the ORU grasping task, more complex operations were performed: for example, failed equipment replacement was achieved as a combination of different grasping tasks. In Figure 5.1.8 some frames show the completion of such an operation.

Simulations without Markers

Within Eurobot programme, ESA confirmed the choice to employ markers in order to simplify the accomplishment of autonomous tasks, considering the high level of reliability thus achieved. However, in order to remove the reliance of

Figure 5.1.7. EGP working set up in Leiden. Courtesy of ESA.

these operations on marker presence, some simulations were carried out and the accomplishment of the same ORU grasping task is depicted in the frame sequence of Figure 5.1.9. This new algorithm, no more needing marker presence, exploits instead the knowledge of the object model.

This could represent a further improvement because it would allow engineers not to count on a rigidly structured environment: in this way the robot could easily accomplish its own operations, even if some kind of error occurred during the mission. Hence, this could be one first step toward a robot endowed with a more "smart" vision system, able to deal with natural features. However, performances in terms of reliability should be evaluated.

Moreover, a situation in which the ORU is not visible from EGP initial position has been simulated too, as shown by Figure 5.1.10. In this simulation, EGP rover has to move until the ORU comes to be partially in view for the camera; as soon as the ORU is visible, EGP vision system recognizes it and can begin its tracking. Both centering and approaching phases toward it are executed, until the handle is grasped by the robot arm.

This simulative results demonstrate that the grasping task can be accomplished even if errors occur, causing the robot not to be exactly where it should be: in fact, object nominal positions are known but EGP could fail its position-

Figure 5.1.8. EGP equipment replacement tested in Leiden, The Netherlands. Courtesy of ESA

ing, for example due to errors and uncertainties in its DEM. Therefore, grasping and manipulation tasks can be completed also in these conditions, thus adding reliability to EGP autonomous operations.

5.1.2. EGP Solar Array Deployment

Another autonomous task to be performed by EGP with the aid of its vision system is the solar array deployment; this panel is located on one side of the Lander, and it has a handle, just like the one on ORUs, as compatible robotic interface

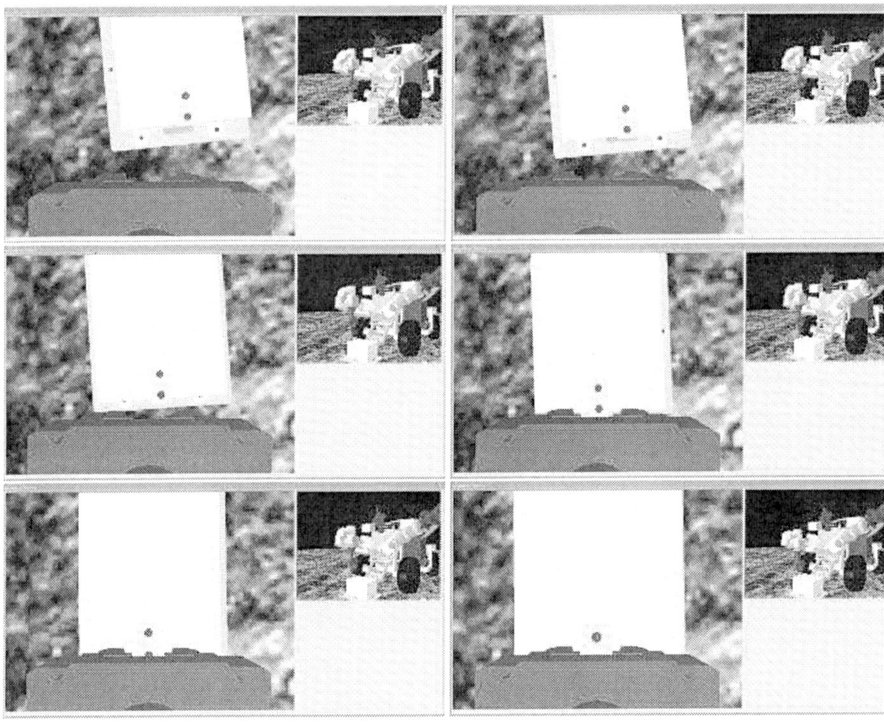

Figure 5.1.9. ORU grasping simulation without marker.

allowing manipulation and grasping operations, as depicted in Figure 5.1.11.

This task, requiring both rover and arm motion, consists in moving toward solar array nominal position and then grasping the handle placed on the solar array panel. After this initial phase, opening of solar array can begin: to this purpose, EGP mobile segment must be able to deploy the solar array along its opening direction, for up to 2 m, as explained by Figure 5.1.12.

At the beginning of the task opening phase, the solar array is assumed to be tightly grasped by EGP; this grasping of the handle mounted on the solar array is accomplished thanks to the same marker-based vision strategy previously described. The robot must then proceed to the deployment itself. The direction, illustrated in Figure 5.1.13, along which the solar array extraction must be executed is parallel to the ground and must follow the intrinsic deployment direction of the panel.

If some errors occur, due to particular initial conditions, and EGP rover is

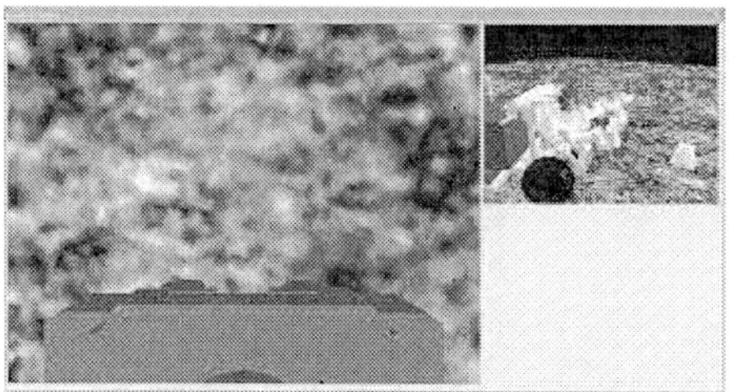

(a) initial position. ORU not visible.

(b) EGP approaching ORU.

(c) ORU in view.

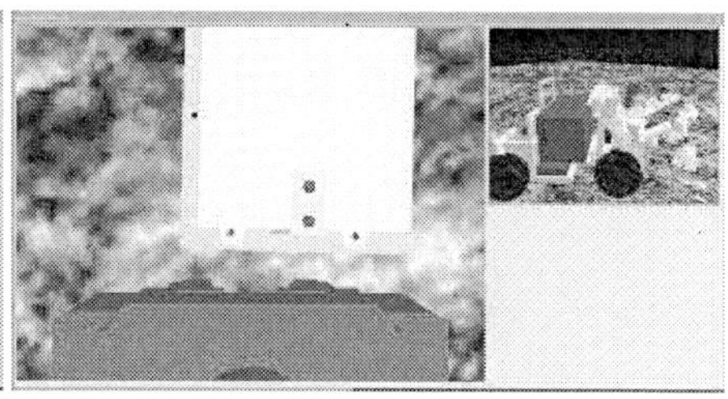

(d) alignment to handle.

(e) approaching handle started.

(f) grasped handle.

Figure 5.1.10. ORU grasping simulation. No marker is used and, at the simulation beginning, the ORU is not visible.

not able to completely open the solar panel, the implemented control law must prevent the robot itself to force solar array to movements that are not along its extraction direction. This must be assured in order to avoid breaking the panel itself.

This task requires a coordinated motion for arm and rover: they both must be controlled and moved at the same time, unless the initial rover positioning error is null. In this last case the arm has to move first, guided by the vision system,

Figure 5.1.11. Solar array deployed from the Lander. Courtesy of ESA.

and then, once grasped the handle, the rover must be controlled in such a way to correctly deploy the solar array, while the arm firmly keeps its grasp. This operation, as well, has to deal with the assumption that the rover positioning is not perfect and EGP initial position for this task is not completely known; the compensation of these potential errors is, as was before, a duty of the vision system that guides the first handle grasping phase.

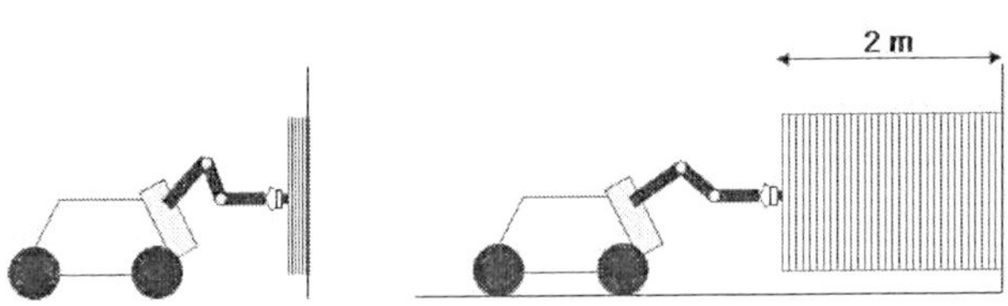

Figure 5.1.12. EGP solar array deployment task.

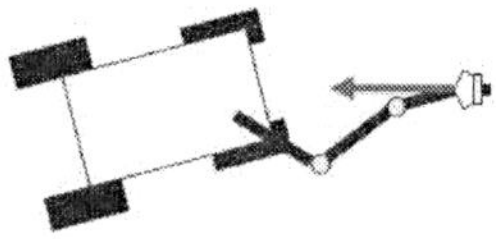

Figure 5.1.13. Solar array extraction direction.

Simulations and Experimental Results

After the vision system has allow EGP to grasp the solar panel, the following deployment phase must begin, along the known extraction direction. At first, a simple control law based on the knowledge of Jacobian matrices of both vehicle and arm was tried; however, this strategy turned out to be bad suited for the task. In fact, when the robot is commanded to deploy the panel, it should align itself to the extraction direction while going backwards; with the Jacobian-based control law, instead, EGP tried to turn forward. This was, obviously, not acceptable as it would cause the robot to break the solar array.

So, a new control law was needed in order to accomplish this deployment task. According to the chosen strategy, a reference command is provided to the arm end-effector and is split up by the algorithm into two motion contributions, one for the rover and the other for the arm.

From the known desired extraction direction d and the known rover longitudinal axis h, it is possible to compute reference speed and jog to be sent to the vehicle control loop. This is done by computing the ϕ angle between rover axis h and direction d vectors as expressed in Equation 5.1.1 (the third component of the vector resulting from the cross product is also checked, in such a way to correctly distinguish the ϕ angle quadrant):

$$\phi = \arcsin \frac{|\mathbf{h} \wedge \mathbf{d}|}{|\mathbf{h}|\,|\mathbf{d}|} \tag{5.1.1}$$

From this angle both linear and angular motion contributions to be executed by the rover can be computed. These contributions, after being processed through rigid body transformations (essential to project the reference itself onto the correct frame), are subtracted from the reference sent to the arm end-effector. Exploiting the extra d.o.f. of EGP arm, a secondary task is also introduced, aimed to maintain the robot posture as much far from singularities and joint limits as possible.

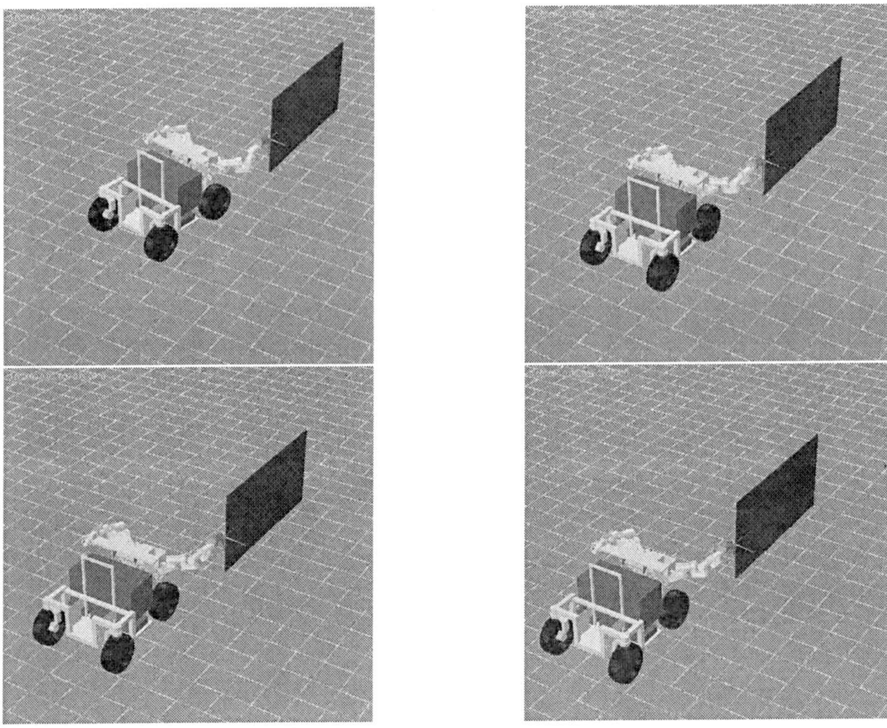

Figure 5.1.14. Solar array deployment simulation. Note that the rover starting position is not exactly along the extraction direction: EGP rover is controlled in such a way to correctly align with the solar panel.

Thanks to this control strategy, both rover and arm are involved in the execution of the solar array deployment and together they succeed in accomplishing this task.

Simulations in Figure 5.1.14 show only the second part of the overall task, the actual solar array deployment. The vehicle starting position is not along the extraction direction: during its backward motion EGP slightly turns in order to correctly align itself to the panel, avoiding potential breaks.

Finally, tests conducted in Dutch Space had confirmed this strategy reliability in solar array deployment for what concerns the experimental phase. EGP, in fact, was able to correctly extract the solar panel along the desired direction and avoiding fatal breaks. Frames taken from this task completion are shown in Figure 5.1.15.

Figure 5.1.15. Solar array deployment execution. Courtesy of ESA.

5.1.3. EGP Cooperative Lifting of Equipment

EGP, as already said, has been studied in such a way to evaluate and demonstrate both automatic and cooperative handling of equipment. Therefore, important functionalities on both autonomy and cooperation side have been implemented and tested in order to let the system work within a test background selected and designed to be representative of real exploration scenarios.

Such challenging tasks include cooperation between the crew on surface and EGP: the robot has to follow astronauts to the work site and assist them during tasks such as moving equipment, multiplying the crew member force capability through its joint actions, while eventually responding to their voice commands. During EGP development, improvements about its speed (but also about different strategies for grasping, vision processing and navigation) were carried out, in order to further increase its usability in the eyes of the human crew during cooperative phases.

Thanks to this design that took into account the user's point of view, EGP is able to physically cooperate with the astronaut e.g. lifting equipment together with him and following the motion of the human while providing strength. A fundamental issue in order to obtain such control objectives is about robot safety while standing side by side with a crew member: EGP behavior must be reliable enough in order for the robot itself not to be a further source of danger to astronauts. This cooperative transportation of equipment can be accomplished through the control and coordination algorithm based on Dynamic Programming and previously described.

To this aim, some preliminary assumptions are to be made; first of all the object is assumed to be firmly grasped by the robot (as previously the grasping phase can be accomplished with the aid of the vision system). In such a way, this object comes to be rigidly connected to the gripper (that is it now belongs to its rigid space) and a tool frame $< e >$ can be conveniently placed at equipment extreme itself. The velocity of the robot end-effector is obtained as the sum of two distinct elements: one for the arm velocity and one for the rover. In fact, if the vehicle is not moving and the only motion is accomplished by the arm, the velocity caused at the end-effector is trivially expressed by the relation

$$v_1 = J_1 \dot{q}_1; \quad J_1 \triangleq {}^0J_{e/b} \tag{5.1.2}$$

being $\dot{q}_1$ the arm joint velocities, J_1 the arm Jacobian matrix and v_1 the cartesian velocity induced to the end-effector by the arm motion only. On the other hand, if the rover is the only one moving, a velocity is imposed by the vehicle to the base of the arm; this cartesian velocity in turn induces to the end-effector a velocity that can be expressed, through the rigid body transformation ${}^0S_{b,e}$, as

$$v_2 = J_2 \dot{q}_2; \quad J_2 \triangleq {}^0S_{b,e} \, {}^0J_{b/0} \tag{5.1.3}$$

being, as before, $\dot{q}_2$ the platform velocity, J_2 the arm Jacobian matrix transformed through the rigid body matrix ${}^0S_{b,e}$ and v_2 the rover cartesian velocity. In case both the arm and the rover are moving, the overall velocity ${}^0v_{e/0}$ induced at the arm end-effector is simply the composition of the two velocity contributions detailed above. Figure 5.1.16 summarizes this described starting situation.

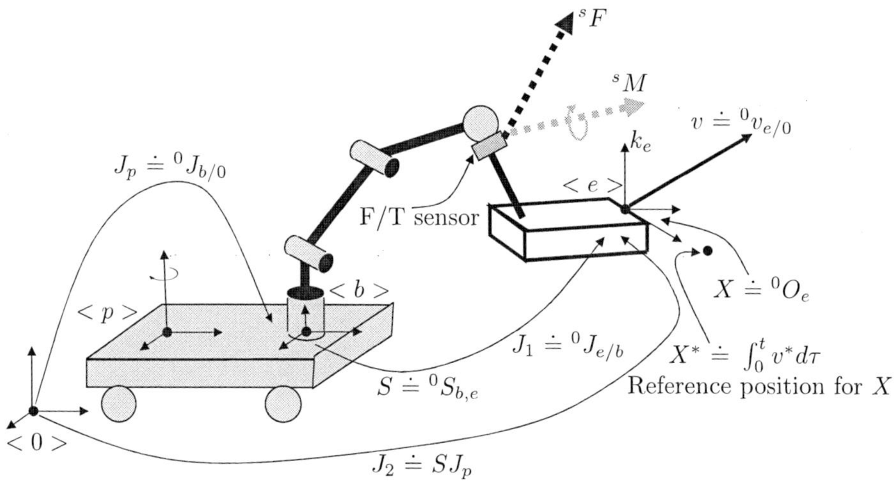

Figure 5.1.16. Frame positioning and main relations among rover and arm.

The general control objectives are basically the following:

- the origin $X \triangleq {}^0O_e$ of frame $< e >$ has to follow the velocity reference command v^*, which is chosen to be proportional to the force vector applied to the robot end-effector and sensed by the force/torque sensor: EGP cooperative task, in fact, is guided by astronaut movements;
- the unit vector ${}^*k_e$ of frame $< e >$ must remain aligned with the unit vector ${}^*k_p$ of frame $< p >$.

The second control objective implicitly corresponds to the apparently natural criterion of keeping the object (that is frame $< e >$) maximally aligned with the slope of the terrain currently being traversed by the rover (and described by frame $< p >$). In the mentioned case, it is requested to keep the equipment somehow horizontal with respect to the rover itself. Obviously enough this command can be changed: for example it is possible to keep the upper object surface orthogonal to the gravity vector (this is the case in which the object to be transported must be maintained upright).

The velocity reference command sent to the overall robot, as already said, is proportional to the force sensed at the end-effector (see Figure 5.1.16); in such a way EGP cooperative task is guided by the astronaut who, applying a force to the carried object, decides which motion direction must be followed during transportation. This type of command potentially implies that a torque ${}^{s}M$ is inadvertently applied to the end-effector by the user: this is not at all a problem since this torque is neglected and only the force ${}^{s}F$ is communicated to the arm controller and interpreted as the velocity command $v^* = k\,{}^{0}F$, upon its projection onto $< 0 >$. Moreover, as in absence of velocity commands the object must maintain its current position, velocity references must be then transformed into desired positions via real-time integration (initializing this procedure with the system initial home position). In this way equipment can remain still whenever the desired velocity reference is zero.

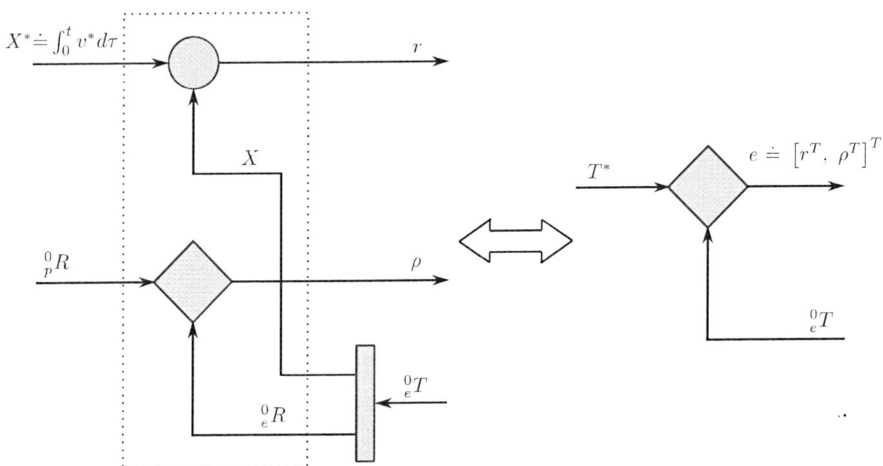

Figure 5.1.17. Error evaluation node.

In Figure 5.1.17 the error evaluation node is depicted, both for linear and angular components. The first one is computed in order to drive the end-effector (and so the grasped object) according to the desired linear velocity. The evaluation error node for the angular component is responsible for aligning frame $< e >$ with $< p >$. Note that, through this kind of computation, the control loop will, or at least will try to, align the entire frame $< e >$ with $< p >$, not only the two unit vectors $^*k_e$ and $^*k_p$. This requirement is more strict than the actual requested one and can be relaxed in order to increase the degrees of freedom that can be used to accomplish secondary tasks, as will be clearer in the following.

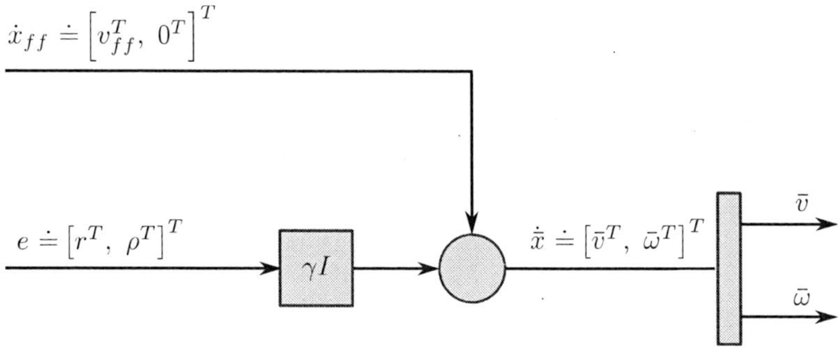

Figure 5.1.18. Kinematic control loop gain.

From Figure 5.1.18, showing the subsequent kinematic loop gain, it is easy to understand the role of each velocity and error vector; the final required cartesian (linear and angular) velocities are represented by $\bar{v}$ and $\bar{\omega}$. These two components can be expressed as

$$\bar{v} \approx v = (v_1 + v_2) = (J_{L1}\dot{q}_1 + J_{L2}\dot{q}_2) \tag{5.1.4a}$$

$$\bar{\omega} \approx G\omega = G(\omega_1 + \omega_2) = (GJ_{A1}\dot{q}_1 + GJ_{A2}\dot{q}_2) \triangleq (\bar{J}_{A1}\dot{q}_1 + \bar{J}_{A2}\dot{q}_2) \tag{5.1.4b}$$

being J_{L1} and J_{L2} the linear parts of, respectively, arm and vehicle Jacobian matrices and, similarly, J_{A1} and J_{A2} their angular subsets. Equation 5.1.4a states that the overall structure accomplished linear velocity v, which is the sum of both rover and arm linear velocities (respectively v_1 and v_2), must be maximally similar to the required linear velocity $\bar{v}$. The same holds for angular velocities (Equation 5.1.4b), with a difference: according to the initial control objectives, only vector $^*k_e$ and $^*k_p$ needed to be aligned by the control law and not all the frame axes, as supposed until now. Hence, a convenient reduction of the angular task is needed, in order to save a degree of freedom and exploit it in a secondary task. This reduction is performed by the G matrix standing out in Equation 5.1.4b and defined as:

$$G \triangleq diag(1,1,0) \tag{5.1.5}$$

The diagonal G matrix multiplies the angular Jacobian of both arm and rover; this means that the orientation about the $^*k_e$ vector is neglected: the only important issue is to keep aligned $^*k_e$ and $^*k_p$, no matter the object attitude about the first one. Note that this is very different from controlling angular velocity about this axis to be null: in the first case a reduction in the Jacobian matrix allows to gain an extra degree of freedom for a potential secondary task, while in the last case no reduction occurs. Considering this reduced Jacobian matrix is equivalent to say that the control algorithm reckons only with the angular error components about vectors lying onto the plane passing through the object upper surface (and in this way $^*k_e$ and $^*k_p$ can be aligned).

Referring back to the Dynamic Programming procedure, the following tasks can be defined, each one with a different level of priority:

$$1)\ \bar{v} \approx (J_{L1}\dot{q}_1 + J_{L2}\dot{q}_2) \ \propto \ {}^sF \tag{5.1.6a}$$

$$2)\ \bar{\omega} \approx (\bar{J}_{A1}\dot{q}_1 + \bar{J}_{A2}\dot{q}_2) \tag{5.1.6b}$$

$$3)\ \bar{\mu} \approx p\dot{q}_1 \quad \leftarrow \quad \bar{\mu} \triangleq \lambda\mu \quad \leftarrow \quad \mu \triangleq \det\left(J_{L1}^T J_{L1}\right) \tag{5.1.6c}$$

Task 1) means that the overall robot must be controlled in such a way to let its velocity maximally approximate to the desired reference one (computed starting from the force exerted by the astronaut). Task 2) is responsible for maintaining the object upper surface orthogonal to the platform z-axis; this $\bar{\omega}$ represents the required angular velocity after the Jacobian reduction described above. Assigning to it a priority 2 decreases the risks of MM losses at arm level: since, in this case, MM is referred to the first task only, the associated arm Jacobian turns out to be "more redundant" than its complete one. In this way, thanks to Task 2) lower priority, the robot has more chances to succeed in following the reference command without significant MM losses. Task 3), as already said, is the safety maintaining task "always do the best for non-decreasing the arm (linear) MM, keeping this measure above an assigned threshold μ_0" (that is in order to maintain a good robot posture). In case MM is close to this threshold, there is the need for this value to be increased; this goal can be achieved imposing a velocity contribution to the robot, in such a way to assure a good posture recovery.

Since Task 3) is at the lowest priority, depending on the required end-effector motion, it may actually occur that the solution of the stated priority task of the control problem cannot eventually guarantee MM above μ_0. This implies that the arm MM must be constantly monitored: this allows the control system to notice if this measure is getting too close to the threshold μ_0. In fact, while trying to fulfill the highest priority tasks, it can occur that MM cannot be kept above its threshold; the more strict these highest tasks, the more frequently this happens. Such a condition highlights the importance of maximally reducing required actions (the role of G matrix, in this case).

Whenever the Manipulability Measure decreases below μ_0, as already explained, a real-time priority change is needed, in order to give this safety maintaining task the highest priority, thus making it become primary.

This described algorithm is currently under validation and its performances have not yet been properly evaluated.

Chapter 6

Stereovision

6.1. Introduction

Stereopsis is at the basis of three-dimensional perception of the world: information about depth can be retrieved from a pair of images coming from cameras that are capturing the environment from a slightly different point of view (as for human eyes). Thanks to this *disparity*, it is possible for human eyes (and so for stereo camera systems) to perceive depth in the surrounding environment.

Hence, computational stereopsis is the process allowing to obtain depth information coming from an image pair and consists of two different steps: correspondence computation and actual perception and reconstruction of 3D world.

For what concerns the first one, it is aimed to find, in the left and right images, pairs of points that are projections of the same three-dimensional points of the scene (those points are named conjugated). This computation is possible thanks to the fact that corresponding images just slightly differ from one another and so left and right projections of the same three-dimensional point are very similar. However, this constraint by itself cannot assure correspondence computation accuracy and thus many false pairs can arise. In order to avoid these uncertainties in correspondence computation, many more constraints can be added, as will soon become clearer.

When corresponding points, relative positioning of cameras and internal parameters of their sensors are known, it is possible to go on with the reconstruction of positions of scene points, which were projected onto the two images. In order to proceed with this step, stereo camera calibration is needed, in such a way to retrieve camera internal parameters (or *intrinsics*) and relative positioning between cameras of the stereo assembly (or *extrinsics*). For details about camera calibration see [27].

6.1.1. Correspondence Computation

As already mentioned, a pair of conjugated points is made up of the projections in the two different images of the same three-dimensional point. The difference between pixels of a conjugate pair, considering left and right images overlapped, is known as *disparity* and is strictly linked to scene depth computation: obtaining disparity values for all the correspondence pairs allows to correctly retrieve depth information.

Several constraints can be exploited in correspondence computation:

- *similarity*: thanks to the slightly different position of the cameras, items in the scene appear to be similar in both images;
- *epipolar geometry*: the correspondent of an image point can be placed in the other image only along the epipolar line that is determined by intrinsic and extrinsic camera parameters (see [37]);
- *continuity*: far from the image borders, depth of points on the same smooth surface changes slowly, thus imposing a limitation on disparity gradient;
- *uniqueness*: one point on the left image corresponds to only one point in the right image and vice versa (this constraint application fails whenever occlusions occur);
- *ordering*: if point m_1 in the first image corresponds to m_1' in the other one, the corresponding of a point m_2 lying on the right (left) of m_1 must be placed on the right (left) of m_1' .

However, even using these constraints, correspondence computation can fail in facing problems such as:

- *false correspondences*: due to the similarity constraint, a point in the first image can be matched with many different points in the second one, thus making identification of correct pairs more difficult;

- *occlusions*: as the cameras have different points of view, some item in the scene are visible only in one of the images, while are occluded in the other. Such points, having no correspondents in the other image, cannot be reconstructed;
- *light distortion*: the more distant the two cameras, the more different the observed light intensity for the same 3D point;
- *projective distortion*: because of perspective, a scene object is projected onto the two images in pretty different ways (this aspect is linked to the distance between cameras, too).

Basically, there are two different classes of algorithm for corresponding point computation: *area-based* and *feature-based* ones. The first kind of technique considers a little area (or window) in an image, searching for the most similar one in the other. This step is done through a correlation measure (or a function of it) for each point in the image. However, these methods have some limitations with respect to uniform regions (for which it is not possible to obtain a disparity value) and gray levels (illumination value in an image is highly dependent on the camera point of view). On the other hand, the feature-based algorithms extract interest points (or features) which have to be as much stable as possible with respect to changes in the point of view. Edges, corners, line segments, regions (or blobs) are some of the possible items that can be considered as features; on these elements the matching procedure is applied, comparing feature attributes among them. These techniques are usually fast but highly dependent on the selected feature extraction method.

6.1.2. Perception of Depth

Considering the simple case in which left and right cameras are parallel and aligned to each other, it is easy to understand that, in such a simple situation (outlined in Figure 6.1.1), the disparity is purely horizontal.

After the world reference frame has been fixed, coinciding to the left camera frame, the following equations can be written:

$$\begin{cases} \frac{f}{z} = \frac{-u_1}{y} \\ \frac{f}{z} = \frac{u_2}{b-y} \end{cases} \tag{6.1.1}$$

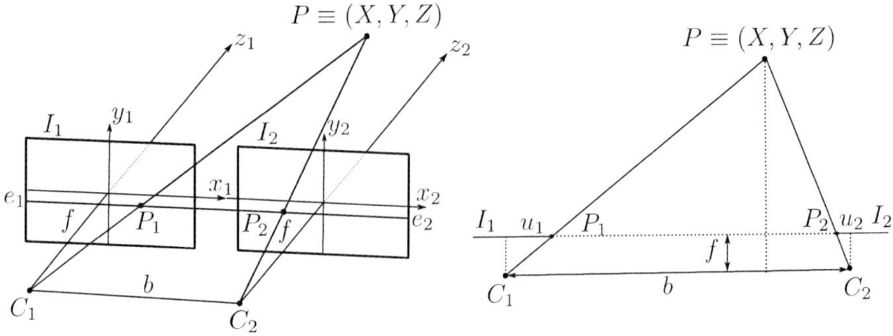

Figure 6.1.1. Stereoscopic rectified reconstruction.

Solving this system, one can obtain the following formula in order to retrieve the depth component of each world point:

$$z = \frac{bf}{u_2 - u_1} \tag{6.1.2}$$

Equation 6.1.2 suggests that the z component can be computed if both system geometry (baseline b and focal length f) and disparity value $(u_2 - u_1)$ are known. The stereo assembly baseline b thus behaves as a scale factor: disparity is proportional to it; moreover, even if b is not known, it is possible to proceed with 3D reconstruction up to a scale factor.

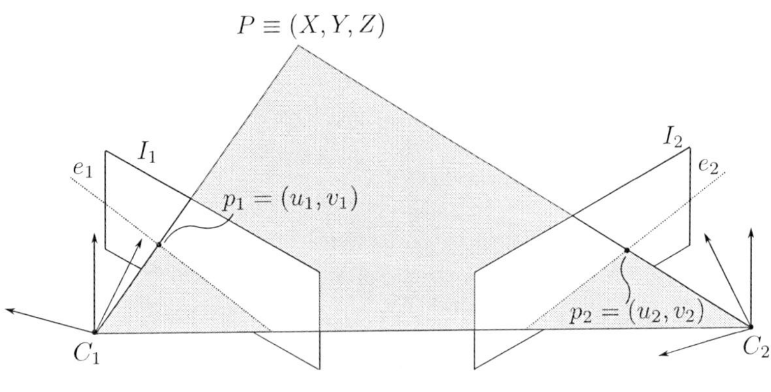

Figure 6.1.2. Stereoscopic reconstruction.

This described situation (aligned and parallel cameras) is a very special and simple case, as usually cameras of a stereo assembly are placed as shown by

Figure 6.1.2; however, by re-projecting the left and right images onto the same plane, it is possible to bring each general situation back to this simple case. This process, known as image rectification (see [37]), consists in applying a linear transformation to each image, thus making them lying on the same image plane. In such a manner, it is possible to simplify 3D reconstruction process, computing depth information thanks to Equations 6.1.1 and 6.1.2.

Nevertheless, depth retrieval from pairs of stereo images is not so easy and robust: small errors in corresponding points, for example due to image noise and not very precise feature detection, highly affect accuracy in 3D reconstruction. For further details about stereovision and depth computation see [37] and [38].

6.1.3. Feature Extraction

The starting point for each 3D reconstruction procedure is the feature extraction step. Several algorithms allow to extract interesting points from an image, depending on different criteria, according to which a feature is classified as interesting by the procedure. If the chosen technique is appropriate, each extracted point encodes a good piece of information, allowing to exhaustively describe the scene with the aid of few features.

Being able to assure a trade-off between accuracy of the described world and resources needed to describe it is a good property for such an algorithm, in order to effectively depict a clear view of the surrounding environment (that is generally complex) with a little computational effort.

Feature extraction algorithms can be subdivided into different groups, according to the particular implementation choice:

- *Low-level procedures*: basically they consist in capturing sharp transitions in image brightness (or, more formally, discontinuities) as they can correspond to changes in the world properties. In this category there are simple techniques such as corner and edge detection, or scale-invariant feature transform (SIFT).

- *Curvature algorithms*: they are mainly aimed to search for changes in image intensity, e.g. across object edges, and find direction of contours.

- *Shape-based techniques*: they proceed to split the image up (strategy also known as segmentation) into different regions, which are distinguished by labels assigned to them according to some criteria. Examples of these algorithms are a simple thresholding, blob extraction and pattern matching. Particular shapes, such as circles or ellipses, can be found through the Hough transform (for details refer to [39] and [40]).

Harris Corners, SIFT and SURF

The most commonly used feature extraction techniques are Harris corner detector [22], Lowe's SIFT [23] and SURF [24]. There are, then, many variations to the previous algorithms and other methods, such as FAST [25] and SUSAN [26], each suitable for the specific application, according to its context and purpose.

Harris corner detector considers the difference of each corner candidate with respect to the main pixel directions: interest points have a good difference towards all of them. This detector valuable quality lies in its computational speed, but it has a big drawback: it is not invariant to scale changes and this can be a problem when moving in an environment, searching for one or more objects.

On the other hand, SIFT features are more and more robust and, thanks to the descriptor assigned to each interest point, this algorithm is able to recognize objects within different images with a greater reliability. In fact this extractor is considered to be invariant to scale, orientation, affine distortion and even partially invariant to illumination changes. Computationally, however, it is slowed down by this descriptor complexity and thus requires a lot of computational resources.

SURF relies on descriptors that are similar to SIFT ones but is faster, even if it allows to effectively track interest points, as it, too, is robust against different image transformations; however its speed is not comparable to Harris corner detector one.

6.1.4. 3D Reconstruction Experiments

Some preliminary tests were performed in order to try to retrieve world depth and locate 3D objects in space (without the aid of any marker), maintaining their relative positioning.

A first set-up was realized with the aid of two GigE (standing for Gigabit Ethernet) Basler scout color cameras, one of which is shown in Figure 6.1.3. They provide a 1024×768 resolution and a fast data transfer rate (up to 1000 Mbit/s, as they exploit the Internet Protocol standard).

Figure 6.1.3. Basler scout GigE camera. Image credits: Edmund Optics.

Nevertheless, in order to use them as a stereoscopic assembly, these cameras were to be arranged together in a precisely known relative positioning and were to be well calibrated too. These procedures were not so easy to be performed with a good precision; small errors in both calibration and relative positioning (intrinsic and extrinsic parameters) of cameras badly affected 3D reconstruction accuracy (above all in terms of object depth retrieval).

Hence, a Point Grey Bumblebee 2 stereo assembly (Figure 6.1.4) was exploited: this system has the great advantages to be already accurately calibrated and to have each camera sensor tightly positioned with respect to the other one.

Thanks to such a stereoscopic assembly, errors due to bad calibration and relative positioning uncertainties were removed and a good depth retrieval was performed. Each Bumblebee 2 camera has a 1024×768 resolution, is full-color and can be controlled through a firewire interface.

A simple algorithm was developed, in such a way to compute depth information and be able to reconstruct the world as captured by the vision system. This procedure is composed by the following steps:

- image acquisition;
- feature extraction;
- feature matching;
- disparity map computation;
- invalid point rejection;

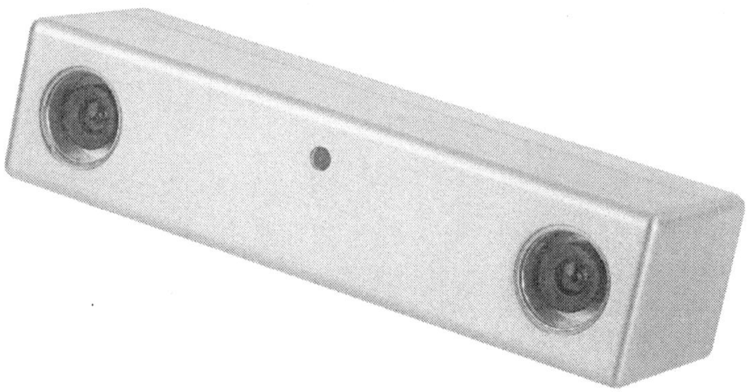

Figure 6.1.4. Bumblebee 2 stereo assembly. Image credits: Point Grey Research.

- triangulation and 3D coordinate computation.

These steps are executed in real-time and, according to the chosen image resolution, a fluent enough flow of images can be obtained. The computed 3D point coordinates are communicated through a simple shared memory mechanism to another simple program, developed using OpenGL functionalities, whose job is that of drawing the 3D points. Frames taken from preliminary tests are shown in Figures 6.1.5 and 6.1.6.

Preliminary experimental results were obtained applying to extracted and matched features only the simple triangulation procedure; this is not sufficient for a good reconstruction: as highlighted by Figures 6.1.5 and 6.1.6, many invalid pixels are kept and included in the final scene displaying. A more convenient strategy for grouping pixels into distinct objects can be used for a clearer reconstruction, thus being able to better represent object shapes. However, even without this more sophisticated technique, experimental results show as object relative positioning is correctly kept during reconstruction process and some objects can still be detected in the displayed scene.

This as much precise as possible 3D reconstruction represents the basis of an accurate rover navigation, in terms of both place identification and obstacle avoidance, assuring also a safe path planning.

(a) Captured left frame.

(b) Reconstructed 3D world displaying.

Figure 6.1.5. 3D Reconstruction of objects on a table.

6.2. Space Applications

6.2.1. Entry Descent and Landing

Recognition of known objects can be useful, in planetary applications: for a robot like EGP itself, a reliable system for object tracking would be a great

(a) Captured left frame.

(b) Reconstructed 3D world displaying.

Figure 6.1.6. 3D Reconstruction of surrounding environment.

improvement, in order to eliminate databases full of correspondences between fiducial markers and objects to be grasped.

Moreover it can be helpful after the launch of any space module. In fact, during the descent toward the planet surface, vision can identify known areas, thus avoiding potential hazards and also approximately locating the module position. This is, for example, the case of the EDL algorithm [41]: such a technique is aimed to help the landing on a planet.

Among EDL purposes, however, the retrieval of system speed and attitude, from changes in the images taken at different time points, is very important in such a way to guide the landing and to be able to compensate, through thruster actions, potential misalignments with the foreseen landing set-up.

The EDL technique has been analyzed for long time by NASA and is being currently studied at Thales Alenia Space for an ESA-funded project named ExoMars, aimed to launch a rover on Mars surface (the launch is scheduled for 2016-2018) with many different scientific and technological purposes.

Atmosphere on Mars is thick for rocket use and thin for parachutes: the first ones cannot be employed to properly decelerate the landing module, while the others, alone, cannot be effective enough in order to significantly affect the descent phase. A possible solution is the use of airbags to reduce the landing impact, but this imposes a very large limitation on the system size. For these reasons, the need of developing a good vision-based technique for EDL arises. Required features for such a system are obviously a high level of reliability and a good accuracy in order to reduce the size of the uncertainty linked to the landing area.

Usually, in fact, this approximate area is known (and then the system will touch down the planet soil within a zone of uncertainty arranged in advance) and also information about its morphological structure is available. Therefore, through these data, it is possible to match known areas of the planet surface with what is seen by the cameras of the system while landing.

This represents another fundamental aspect as it allows to detect potential hazards and avoid them: for example, if the vision algorithm recognizes a crater below the landing module, it may prevent the system from landing inside it, event that would probably cause the mission failure. Similarly, preventing the module from landing on the top of a high crest is essential in order not to have the landed system (and so the potential rover, too, like in the ExoMars case) blocked and unable to perform its scientific exploration tasks, thus becoming completely useless.

Figure 6.2.1. Mars Emulation Environment, at Thales Alenia Space, Turin. Courtesy of Thales Alenia Space.

Hence, preliminary tests have been performed, in order to develop a simple algorithm able to recognize different terrain features. To this aim, a three-dimensional plastic available at Thales Alenia Space has been used as a testbed. It is called *Mars Emulation Environment* and it is depicted in Figure 6.2.1. It is representative of different kind of terrain features present on Mars: there are both sandy and rocky soil features (in the lower band, respectively on the right and on the left), a rift, a crest and a crater (in the middle band, moving from left to right) and, finally, a sort of plateau and a peak in the upper band (moving again from left to right).

Thanks to this three-dimensional Mars emulator and with the aid of a KUKA arm it was possible to simulate different trajectories approaching Mars terrain. Along these trajectories, emulating the space module descent, the vision system can be tested, with different purposes.

Figures 6.2.2, 6.2.3 and 6.2.4 show as different kind of terrain features can be identified; in these tests the system only searched for one feature at each time (represented in the left-upper part of frames in Figures 6.2.2, 6.2.3 and 6.2.4), but this case can be simply generalized to the contemporary search for many different features.

This preliminary algorithm is based on a SURF feature extractor and interest point matching is accomplished on the basis of each feature descriptor only. During the space module descent phase, it is possible to exploit the computed speed information of the module itself, thus improving matching accuracy.

For vision-based EDL purposes, good features to be tracked and identified are craters (nowadays many craters and their morphology are known, both on Mars and on other surfaces such as the Moon, and have even a proper name). Scientific literature dealt with crater matching procedure exhaustively enough (see, for instance, [28] and [29]).

After the landing phase, exploration tasks and scientific experiments, like in ExoMars programme, usually need to be conducted by a mobile robot, able to perform many operations by itself. For this kind of rover, navigation capability is fundamental: making these rovers able to go around the planet surface and to find interesting places for scientific research is a long term objective of space robotics. This goal accomplishment would drastically decrease the need of human presence in such missions, reducing also the problems linked to it.

To this aim, robotic vision is an essential ability to develop, as it can help rovers to identify interesting planet areas according to some criteria, to navigate avoiding obstacles and preventing damages and also to compute, from image changes, their own motion.

This last issue is known as *visual odometry* and represents a good option as an alternative to classical mechanical odometry, as this last can be easily deceived by non-smooth terrains being traversed.

6.2.2. Rover Navigation and Hazard Detection

The first basic capability of a rover navigation control system is that of assuring execution of safe paths only, e.g. avoiding impacts against external objects or planetary surface features (for instance a large rock), eventually computing an alternative path to move around these obstacles.

Vision can play again a fundamental role in satisfying this need: thanks to stereo cameras devoted to navigation, DEM maps can be built, gathering information about depth from 3D reconstruction procedures (after all human beings, too, can avoid obstacles thanks to their eyes).

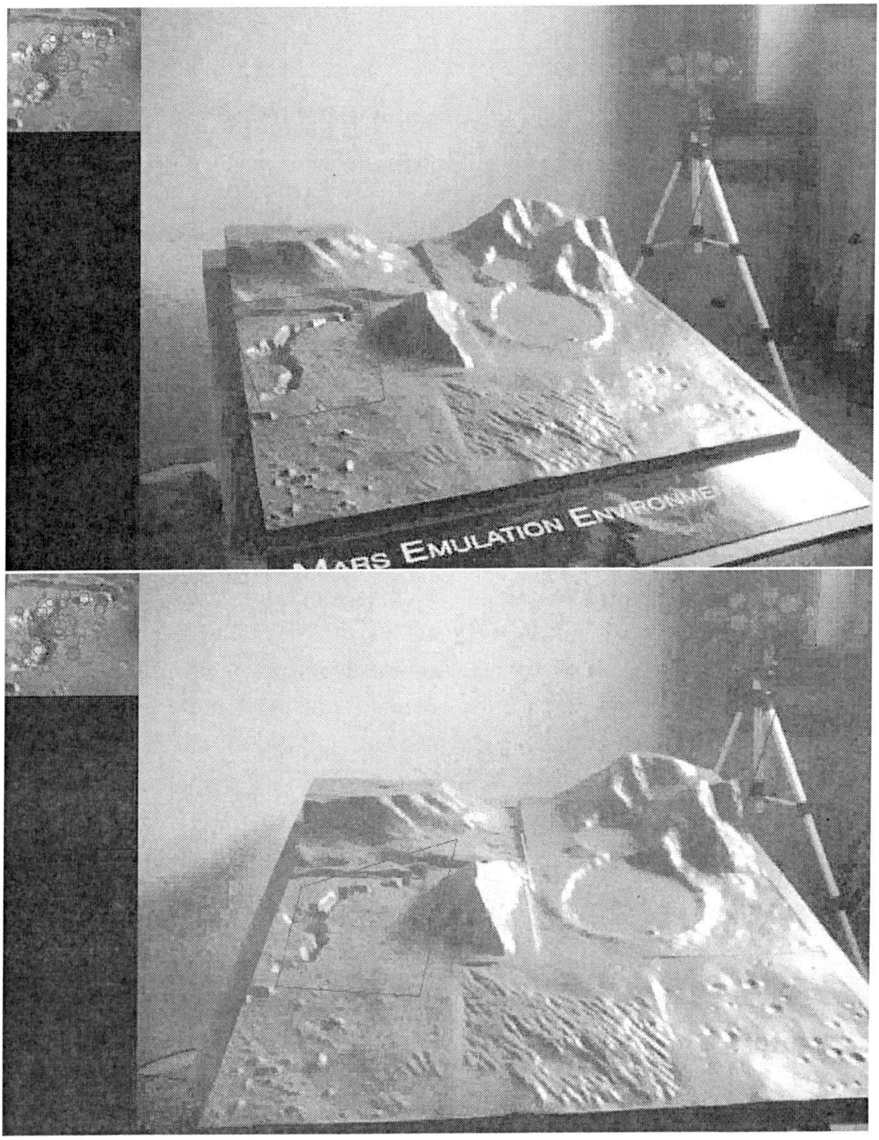

Figure 6.2.2. Tracking a rift on the Mars Emulation Environment, at Thales Alenia Space, Turin. Courtesy of Thales Alenia Space.

ExoMars, as an example, exploits a stereovision software developed by

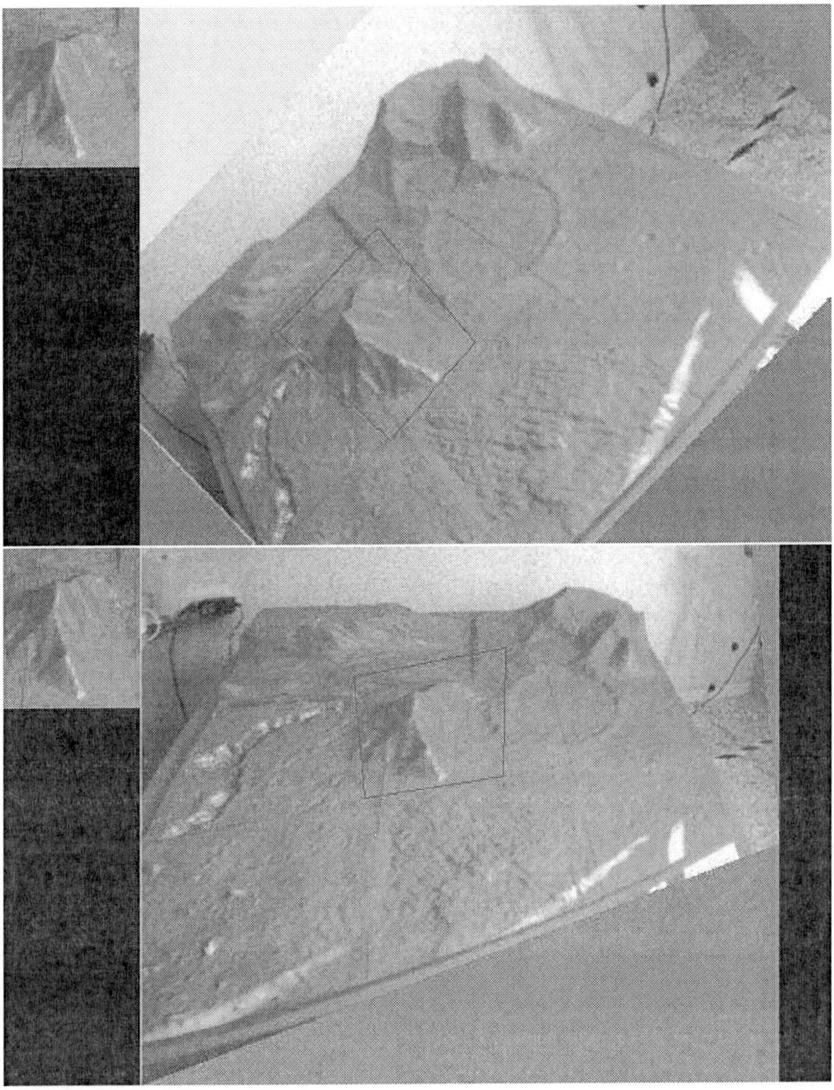

Figure 6.2.3. Tracking a crest on the Mars Emulation Environment, at Thales Alenia Space, Turin. Courtesy of Thales Alenia Space.

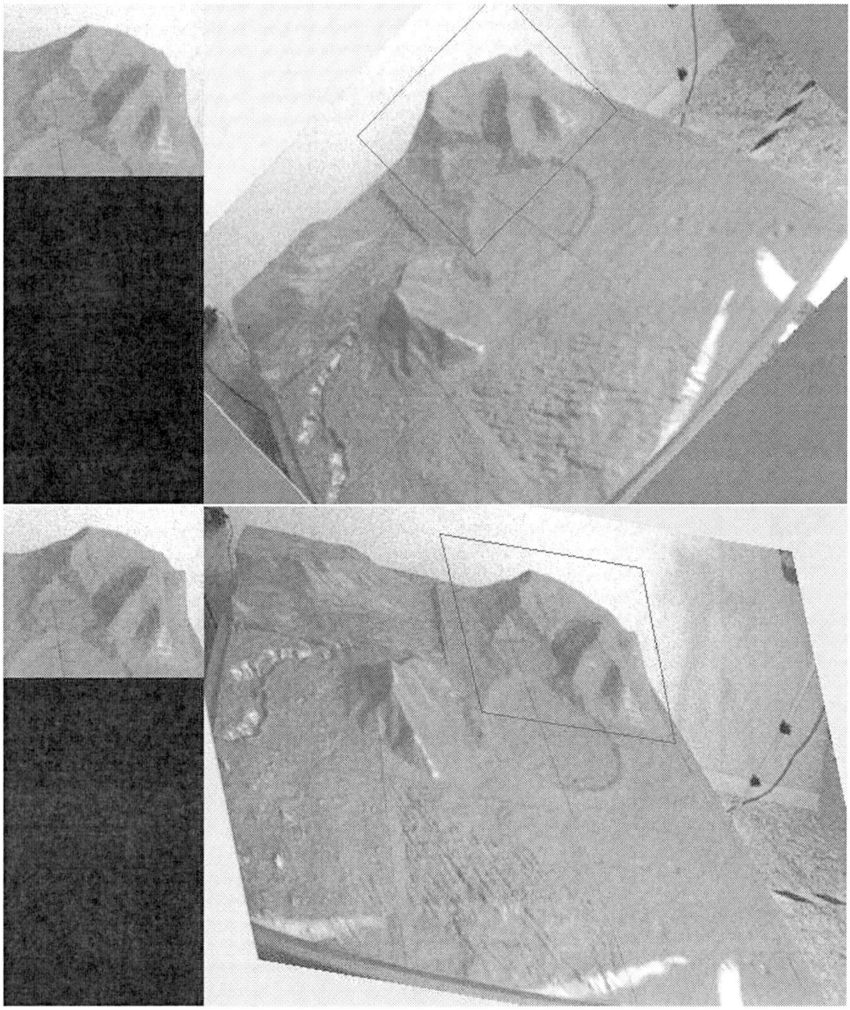

Figure 6.2.4. Tracking a peak on the Mars Emulation Environment, at Thales Alenia Space, Turin. Courtesy of Thales Alenia Space.

CNES [30]. This algorithm allows to three-dimensionally reconstruct surrounding environment morphology, starting from two images of the scene as perceived by a stereo camera assembly.

Basically this technique is made up of the following procedures:

- Correlation-based matching: through a simple correlation between the two image luminance functions, for every pixel of the first image, a correspondence in the second one is searched for. A score is assigned to each potential point pair (matched between left and right frames).

- Disparity map computation: for each pair of stereo frames a disparity map is built up, containing all the needed data for 3D reconstruction.
- Disparity map filtering: this step helps to decrease mismatches (caused by the previous correlation computation), above all happening when several pixels have similar correlation scores.
- Stereovision: for each pixel, among the computed scores assigned to it, the match with the higher score is chosen as a valid pair and the corresponding 3D point is reconstructed.

This algorithm, actually, is applied to a 2-level pyramid of images in order to increase speed and reduce mismatching. To this aim, an image degradation is performed: for smaller images the required computation is much reduced. In fact, for this technique complexity the following expression holds

$$C \propto X \cdot Y \cdot D_{disp} \tag{6.2.1}$$

where X and Y are respectively width and height of the input images, where as D_{disp} is the disparity interval provided by each camera (this range highly depends on its intrinsic parameters, such as resolution, stereoscopic baseline, focal length and so on). Reducing image resolution by a factor r, the new obtained complexity is

$$C_r \propto \frac{C}{r^3} \tag{6.2.2}$$

Hence, on these smaller images, required computational resources are highly lowered.

Degraded left and right images come from acquisition devices that are different (for example because of their CCDs and lens systems); thus these frames need to be normalized with respect to light, in order to reduce errors in the following reconstruction procedure.

The previously detailed algorithm steps are performed on both the original and the degraded images; this multi-level pyramid method allows to highly reduce the search windows in the stereovision algorithm, thus decreasing the required time for computation too.

Retrieving the correct disparity map (note that only horizontal disparity is considered, as camera optical axes are regarded to be parallel by construction or rectification) corresponds to finding the correct matching points between left and right images. This goal is accomplished through the correlation step: for each point P_1 in the left frame, through a search window with a predefined size, a corresponding P_2 in the right image is selected. This choice is made on the basis of scores assigned to each potential pair $P_1 - P_2$: the corresponding point chosen as right matching of P_1 will maximize this score in a proper range depending on d_{min} and d_{max}, respectively minimum and maximum disparity values. This is not enough to effectively validate the retrieved matching pairs and not to be deceived by errors due to occlusions and lacks of matching points in some image areas; that's why the back correlation is computed. It simply consists in applying the same above described procedure, this time starting from P_2 in the right image and searching for a corresponding match P_1 in the left frame. Only pairs of double-matched points are kept as valid and their 3D coordinates are computed, using a simple triangulation method (Equation 6.1.2). Figure 6.2.5 shows an example of a possible terrain and its resulting disparity map.

Disparity map filtering is a key issue as it prevents mismatches that can lead to the computation of a wrong distance. The corresponding cell in the DEM is often interpreted as an obstacle in the further steps of the navigation algorithm; hence the importance of a good filtering technique is emphasized. Isolated errors are eliminated thanks to topological considerations, assuming that the terrain is locally a continuous surface: through image segmentation, connected areas are retrieved and a filtering is performed, aiming to remove regions that are "floating in space" and that have a size below a defined threshold. Figure 6.2.6 illustrates how the disparity map filtering affects the resulting DEM.

At the end of this algorithm the data are structured as a set of 3D points, each corresponding to a specific image pixel. In order to make terrain analysis easier, these points are projected onto a regular horizontal grid with fixed dimensions and an altitude is associated to each one of them. The several points that are projected onto the same cell are not averaged: the minimum and maximum values are retained in such a way to provide a "worst case" estimation of the obstacle altitude (in order to guarantee a safer path planning). DEM, computed with a very high resolution with respect to wheel size, can be displayed as a continuous surface with the addition of a simple texture or shade, as in Figure 6.2.7.

After the DEM has been built up, a navigation map is needed in order to mark out non navigable areas. This is done determining, for each pixel of the

(a) Left image after distortion compensation.

(b) Disparity map before filtering.

Figure 6.2.5. Left image from one of the cameras and computed disparity map. Courtesy of CNES.

DEM, whether the rover is capable to cross the corresponding cell or not; cells that can be traversed are assigned to a difficulty score, according to their level of ease in navigability. An example of navigation map, corresponding to a single stereo pair, is depicted in Figure 6.2.8, where the rover is placed at the map center. Black means unknown areas, white corresponds to forbidden ones, while the gray underlines navigable zones, with a different intensity according to the difficulty score.

A single perception and the concerning DEM are not enough to manage path planning on long distances: a feasible trajectory cannot be found if obstacles are detected within the small local region analyzed because the rover cannot avoid entering dead end areas. Hence, to overcome this problem, each current perception is merged with the previous ones, in order to update the global navigation map (Figure 6.2.9); the most recent perception has higher priority in situation of overlapping regions.

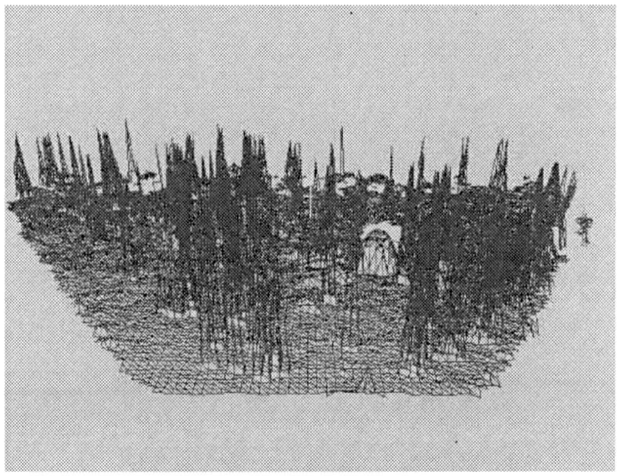

(a) Terrain model before filtering.

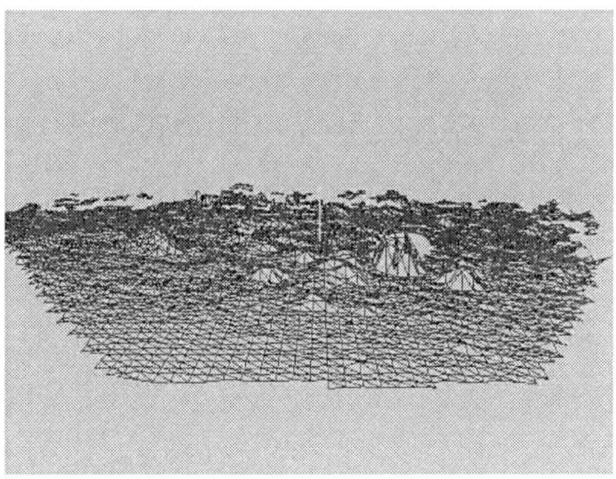

(b) Terrain model after filtering.

Figure 6.2.6. Effect of filtering on the DEM. Courtesy of CNES.

An essential issue to be considered is represented by both the scarce accuracy of rover trajectory execution and the errors of its localization system. These aspects are taken into account by increasing the size of the obstacles by a radius that is proportional to the maximum error expected. As this method is iterative, obstacle size keeps growing until a new perception is performed, updating their dimensions. This increasing growth of obstacle size, however, causes a situation in which the robot is considered to be stuck within non navigable areas: due to obstacle enlargement no path to the goal can be retrieved, even if this is not true, as shown by Figure 6.2.10. To overcome such a problem a fixed value can be used to increase obstacle size.

(a) Shaded DEM.

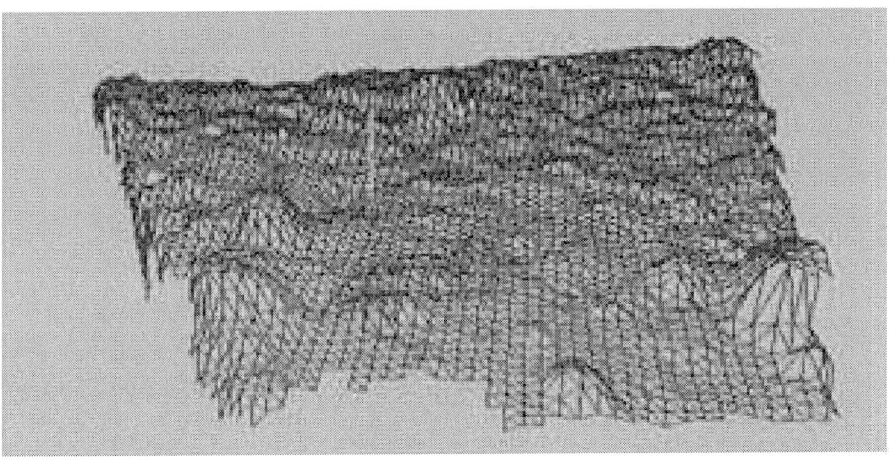

(b) Wire DEM.

Figure 6.2.7. Different representations of the same DEM. Courtesy of CNES.

After the navigation map has been obtained, the next path planning phase begins: given a target position, the rover must move toward it, within navigable areas. This goal is accomplished by selecting intermediate sub-goals according to the following criteria:

- each of them should be as close as possible to the final target position;

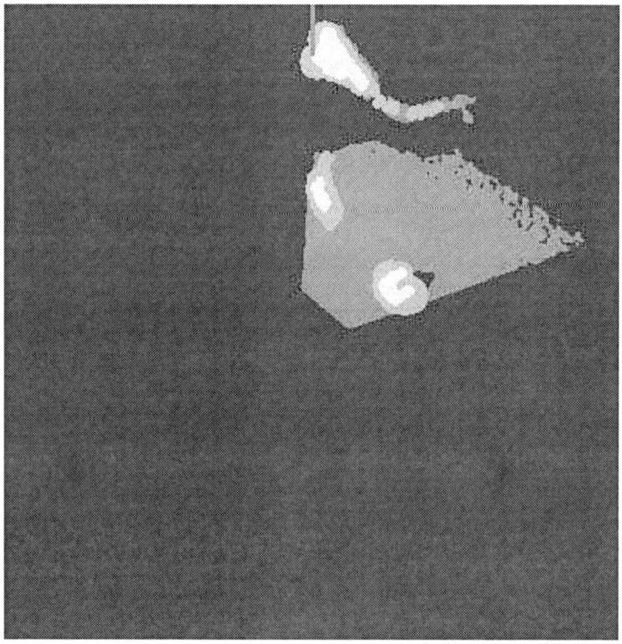

Figure 6.2.8. Navigation map from a single stereo pair. Courtesy of CNES.

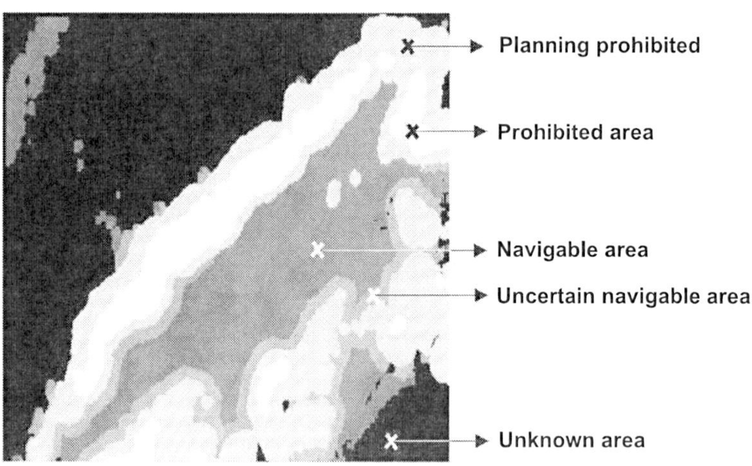

Figure 6.2.9. Global navigation map with different areas. Courtesy of CNES.

Figure 6.2.10. False rover dead end due to obstacle size growing. Courtesy of CNES.

- they should not be at non-navigable area borders;
- each of them should be placed within a navigable area continuously connected to the rover location.

An optimal path toward the sub-goal is then computed, keeping into account both difficulty scores and distance; this last is only partially considered because the overall trajectory is not executed. In fact the planning should be able to let the rover avoid dead end situations and, eventually, come back along its track when necessary. Moreover, trajectories need to include the safety margins and, as these ones grow with the increasing distance, it is possible that, considering a long path, a potential trajectory is ignored as being closed by these safety margins. So, a portion of this path is executed (Figure 6.2.11) and safety margins are applied to it only. Hence, path planning is performed in the following steps: a path (not necessarily optimal) to the sub-goal is firstly found and approximated by a segment "skeleton"; a short term goal is selected along this skeleton; finally, an optimal path is computed toward this short term goal. All these phases are shown in Figure 6.2.11

The found trajectory is then sent to the rover control loop and executed. At the end of the rover motion, its position is almost known; an essential issue to the prosecution of its traverse is the known area extension. In fact, determining the direction of the stereo camera line of sight that maximizes terrain knowledge,

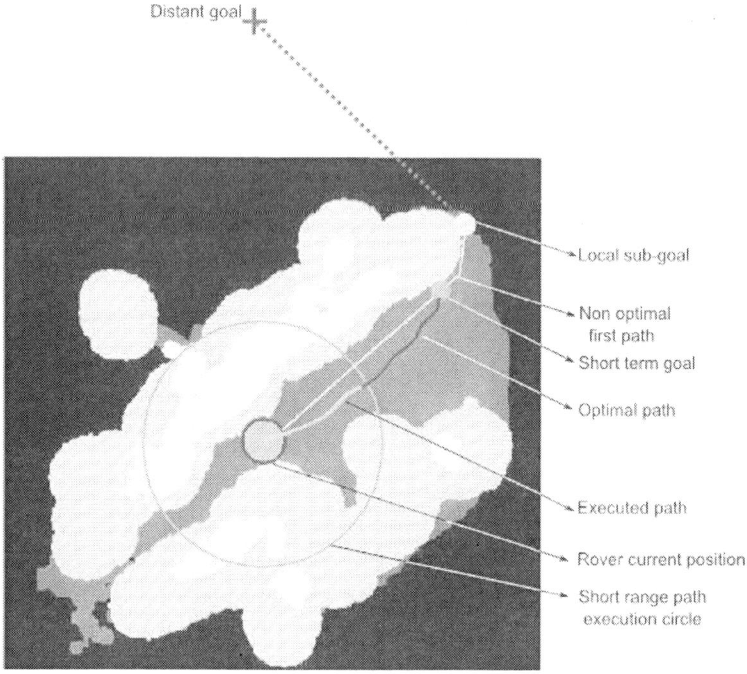

Figure 6.2.11. Path planning phases. Courtesy of CNES.

it is possible to observe useful regions (e.g. toward the final goal and around non navigable zones). Figure 6.2.12 shows the determination of future camera views.

All these steps are repeated until the rover reaches the goal position or until the path planning fails because of a dead end situation, in which the rover is considered stuck in non navigable areas. An outstanding problem is, however, the execution precision of computed paths: an effective solution is required in order to accurately measure the robot motion. In fact, both relative and absolute position knowledge is fundamental in such a way to control rover locomotion and navigation, allowing the control system to detect slipping situations, to correctly reconstruct DEM and 3D surrounding environment, and to reach final goals and execute scientific tasks. Rover attitude and position are usually retrieved from accelerometers and wheel odometry; both of these methods are subject to a great amount of error, due to non-smooth traversed terrains and to long traveled distances. Moreover, also absolute rover position, linked to higher

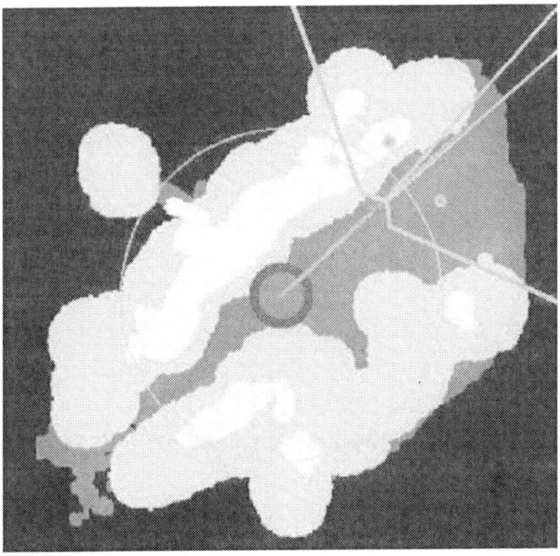

Figure 6.2.12. Useful camera directions for a new perception cycle. Courtesy of CNES.

mission purposes (like scientific experiments to be conducted in specific planet regions) and which is obtained from orbiter data, is subject to errors from 10 to 100 m. Hence, the need of a different method to compute rover motion: visual odometry is one of the most promising technique to be used, at least to complement classical odometry.

6.2.3. Visual Odometry

Visual odometry, working side by side with a navigation and obstacle avoidance system, can be a fundamental solution to the rover motion estimation problem; this is because it is not affected by rugged terrains and by situations such as slippage (this is very important because hazard detection and obstacle avoidance highly rely on position estimation and wheel odometry cannot be accurate enough). Moreover, it does not need further sensors, as it uses stereo cameras, already planned to be mounted on board any rover.

For all these reasons, there is a great interest in an effective accomplishment of this motion measurements, observing changes in the stereo pairs. State of the art for visual odometry applied to space applications is described by NASA within MER project first [31] and MSL then [32].

MSL visual odometry, more recent and effective than MER one, is based on a 4-level pyramid: for each stereo pair a downsampled set of images is built up and the algorithm starts from the smaller left frames. The Harris corner detector [22] is used, in order to retrieve many features to track; these features are then matched against the right smaller image through a pseudo normalized cross correlation step [33], exploiting the epipolar geometry constraint and considering a limited disparity range. This step ends with the knowledge of 3D points, triangulated starting from the matched features (on the more coarse level image). Thanks to these points and using the initial motion estimation (for instance coming from wheel odometry), the features corresponding to the same reconstructed 3D points are searched for in the next (in terms of time) left frame; these features are then matched with the right corresponding image. Through another triangulation step, a new set of 3D points after motion is provided. At this point an outlier rejection algorithm is applied, based on a so called rigidity test: distances between all points before motion is computed and the same is done for points after motion; only points with similar distances from the others before and after motion are kept. After this rigidity test, a RANSAC algorithm step [43] is performed and an estimation is retrieved and employed as initialization to the next phase. This first estimation is obtained considering several subsets of corresponding points and computing the related transformation matrix $[R|t]$; according to the distance among extracted and re-projected (through each matrix $[R|t]$) points, scores are assigned to every computed $[R|t]$. The best one (having the minimum distance) is used as initialization for the following iterative step, the maximum likelihood estimation (MLE) [42]. At the end of this phase, the final motion estimation (given to the next level as an initialization) is computed, together with its covariance. This last is used to limit the search windows in the next level: through the computation of the covariance matrix SVD, it is possible to compute, for each point, a bounding box; this is done in order to limit the search of features in the next level corresponding to the 3D points previously extracted. All these steps are repeated for each pyramid level and for each captured stereo pair.

The multi-level pyramid slows down the computation because the same steps must be repeated on more images; however, thanks to the covariance decomposition, the search window size is highly reduced and so a few computational resources are needed (only for the first level of each stereo pair covariance is large, but the image itself is small and so few time is required for computation). Further attention must be paid to the number of tracked features: problems can arise, in terms of precision, if their number is too small.

Possible Improvements

This algorithm is robust enough but is computationally heavy and so cannot be effective for real-time motion feedback estimation. One of the proposed improvements consists in extracting features from both images of each stereo pair (see [44]) and performing correlation steps only among the detected interest points. In such a way matching step is faster, but the doubled extraction make computational time increase and the feature quality significantly decreases.

Focusing attention on feature extraction procedure, a more uniform feature detection can be performed, thus obtaining interest points more spread in the images; this leads to an improvement in motion estimation quality. Furthermore, introducing a more strict hypothesis of smooth terrain crossing, a variable range of disparity value can be adopted: according to the position in the image of a specific point, its right and next in time matching can be searched for with a specific smaller disparity range (features that are far away move in the images much less than features that are close).

Little by little, errors in motion estimation increase; to reduce this drift, a possibility is that of execute fast estimations of short movements and then add them all up. After this, a complete algorithm cycle is performed between the first and the last stereo pair of the overall motion, using as initialization the sum of the small motion estimations. This algorithm does not rely on wheel odometry but, in order for it to correctly work, the rover must move slowly and the extraction procedure must be fast. Estimation independence from wheel odometry is an essential achievement: in situations of rover slipping on the terrain, classical odometry goes wrong and so badly affects motion visual estimation.

Harris corner detector can be replaced by other algorithms, such as SURF: this technique surely is heavier but it is also more robust and, as there is no more the need of a multi-level image pyramid, the overall algorithm is faster than the previous one. This is also due to the type of interest point matching: features are coupled according to their descriptors and correlation is no more used. Preliminary tests on this SURF based procedure highlighted a good behavior if motion between two following stereo pairs is about $10-20$ cm (error less than 0.5 cm). For stereo pairs with a 640×480 resolution an estimation is obtained, after a complete algorithm cycle, within $600-800$ ms.

Figure 6.2.13. Terrain feature recognition in the simple terrain generation simulator.

Other feature tracking techniques, such as optical flow, can be exploited to further reduce computational time. This last idea, still under investigation, allows to obtain a good enough estimation within a very low time (and so within few changes between stereo pairs); after some tracking, the retrieved overall estimation can thus be used as initialization of an estimation step between the first and the last stereo pair of the entire motion. For this last process, an heavier algorithm can be used, as the optical flow provides a good initial estimation. When both algorithm estimations are available, a weighted mean value of them (according to some criteria) can be provided as motion feedback to the rover control system.

Moreover, retrieving the platform absolute position can be essential with respect to high level scientific purposes: for instance the rover can be required to recognize a determined place, interesting from the scientific point of view. In order to try such a situation, a simple terrain generation simulator has been developed, aiming to represent a planetary surface reproducing Mars soil. The same algorithm used for EDL was employed and turned out to work well on these images, too, as highlighted by Figure 6.2.13: the rover vision system succeeded to correctly identify the desired location.

Chapter 7

Conclusions and Future Research Activities

The main goal of space robotic systems is that of provide a valuable support to astronauts, because work in space is hard and often dangerous to man; moreover, thanks to their current level of autonomy, space systems can save valuable astronaut time by relieving the human crew of many operations and executing such activities in a completely autonomous manner.

Many different approaches to develop effective robotic skills for space exploration were described. All of these methods were aimed to improve mission quality: the achievement of intelligent autonomous systems that could actively and reliably perceive surrounding environment and act accordingly to each potential situation is a fundamental goal for future planetary exploration applications.

Several topics were presented in this Thesis, regarding different but equivalently important aspects. For what concerns space robotics, one of the major interests is that of develop effective techniques in such a way to allow automatic system to interact with their working environment in a safer and more efficient manner. In this context, vision surely plays a fundamental role, as it potentially is one of the most effective sources of information to be employed in space applications.

Vision systems, however, are not yet ready to possess all the desirable features requested to them. Proof of this fact is EGP vision system itself: in order to achieved the high level of reliability required to the robot, ESA chose to use markers assigned to each object of interest. The same EGP operations accomplished only with natural feature extraction (i.e. without the constraint of working in a partially structured environment - at least for what concerns objects on the Lander) would be characterized by a much lower level of reliability. This last feature, obviously, cannot be quit by space missions: once on the planet surface, the robot can be controlled by ground operator or by astronauts in a very limited manner and so few interventions can be carried out to avoid the risk for a mission failure.

Speaking about EGP, good interaction skills and environment perceiving capabilities were achieved; this features, however, are not independent from this specific application and rely on a partial knowledge of the environment (remember marker presence).

While, on one hand, this increases EGP reliability, on the other side it restricts its area of applicability: such a robot cannot yet be employed, for instance, with the aim to recognize, all by itself, a planetary surface region with some scientific interests and to take samples. This kind of operations will be possible only after the developed vision system is completely freed from any marker presence: many research activities about the correct processing of natural features extracted from the surrounding environment must be carried out, so that the robot can obtain useful information about environment structure, thus acting accordingly to it.

Furthermore, for what concerns robotic coordination, a suitable control algorithm has been developed; its positive aspect consists in the possibility to employ it in different situations, no matter the structure of the robot to be controlled. Moreover, as already explained, the definition of several control objectives is possible and easy to implement. Within the framework of Eurobot programme, performances of this coordination algorithm were not yet properly evaluated; future work will involve also this assessment about its effectiveness while performing complex operations and assuring the fulfillment of a great number of control objectives.

Additional future possible work concerning EGP is about a future potential flight model: in the event that ESA resolves to realize, many problems arise. In fact, besides the same already considered needs to have an effective and reliable working system, the flight model would also be bound to the use of obsolete hardware components. In particular, for a control strategy strongly relying on vision, this can be a big problem, in terms of both image resolution and required computational time. These two aspects, in fact, are strictly linked each other: the higher camera resolution (and so image size), the slower the image processing system. In this case, a trade-off between performances and quality of the image processing should be accepted.

Preliminary considerations and results about visual odometry have been presented too; this valuable robotic skill represents one of the steps toward the accomplishment of a more natural interaction with the surrounding planetary surface, without relying on a partially known structured environment. In fact, visual odometry is aimed to retrieve the three-dimensional world shape starting from a suitable feature extraction; depth information is retrieved with the aid of stereovision and no other assumption are needed to proceed with this technique. During the preliminary tests, velocity requirement has been favored, since visual odometry is employed in rover navigation and must be reactive with respect to changes in the external environment. However, reliability is again a requirement that cannot be renounced, otherwise many risks of rover navigation failure arise, thus compromising a potential exploration mission. Preliminary results seemed to achieve a good enough precision in three-dimensional world reconstruction; however, a more accurate evaluation of performances must be executed, as well as an assessment about the level of reliability of this system. Visual odometry development is fundamental in order to endow space robots with the capability to efficiently interact with hostile environments and traverse hard and unknown terrains; in order to obtain effective and fast systems, great work is needed, above all about efficient techniques to extract information from captured world images.

Finally, in anticipation of future missions to the Moon and Mars, the so called EDL technique assumes a fundamental meaning: many studies are aimed to develop robust algorithms in order to effectively assist the landing on a surface. EDL main objectives are the estimation of both speed and attitude of the landing module, in order to drive the module itself and compensate (e.g. through thruster action) for its potential configuration errors. Moreover, while landing, great importance is given to pattern recognition: since many morphological elements of both Moon and Mars surfaces are well-known, recognition and matching of known regions that are being seen by the vision system is fundamental in order to identify possible dangerous situations (for instance if the module is landing within a deep crater or on the top of a high crest) and avoid them. Several studies on EDL are being carried out at Thales Alenia Space for both Moon and Mars surfaces (few differences can be listed between these two cases); again, many problems in the accomplishment of a robust algorithm arise from the employed hardware. In fact, as the landing phase is very fast, without good computational resources it is very difficult to take a enough images to compute speed and attitude and to perform pattern matching. In this sense, research activities on how to speed up these operations are being conducted at Thales Alenia Space; in particular, employment of parallel FPGAs (obviously only space qualified ones) to split computation up between different processors is being analyzed.

Different fields of application were faced in the development of this Thesis; their common objectives are aimed to overcome all the challenges arising from the great interest in planetary exploration and from the future space missions (to the Moon and Mars) that have to strongly rely on a robotic presence. Many research activities and new technique development are greatly needed so that roles of increasingly importance could be assigned to space robots, thus relieving astronauts from dangerous or annoying duties and enhancing the possibilities to accomplish more and more demanding missions.

References

[1] T. Kröger, D. Kubus and F. M. Wahl, "6D Force and Acceleration Sensor Fusion for Compliant Manipulation Control", *IEEE/RSJ International Conference on Intelligent Robots and Systems*, Beijing, China, October 2006.

[2] B. J. Waibel and H. Kazerooni, "Theory and Experiments on the Stability of Robot Compliance Control", *IEEE Transactions on Robotics and Automation*, February 1991, vol. 7, no. 1, pp. 95-104.

[3] G. Bradski and A. Kaehler, *"Learning OpenCV: Computer Vision with the OpenCV Library"*, O'Reilly.

[4] C. P. Lu, G. D. Hager and E. Mjolsness "Fast and Globally Convergent Pose Estimation From Video Images", *IEEE Transactions on Pattern Analysis and Machine Intelligence*, June 2000, vol. 22, no. 6, pp. 610-622.

[5] B. Kainz and M. Streit, *"How to Write an Application with Studierstube 4.0"*, Technical report, Graz University of Technology, 2006.

[6] J. Cai, *"Seminar Report: Augmented Reality: the Studierstube Project"*, Seminar report.

[7] D. Wagner and D. Schmalstieg, "ARToolKitPlus for Pose Tracking on Mobile Devices", *Proceedings of the 12th Computer Vision Winter Workshop* (CVWW'07), Sankt Lambrecht, Austria, February 2007.

[8] L. Kerstein, F. Steinsiek, G. Degtyarenko, D. Brockschmidt and E. Graf, "Inspector and the international space station", *Space Technology*, March 1996, vol. 16, no. 2, pp. 83-96.

[9] J. L. Bresina, M. G. Bualat, L. J. Edwards, R. M. Washington and A. R. Wright, "K9 Operation in May Š00 Dual-Rover Field Experiment", *Proceedings of the International Symposium on Artificial Intelligence, Robotics and Automation in Space*, Montreal, Canada, June 2001.

[10] T. L. Huntsberger, H. Aghazarian, Y. Cheng, E. T. Baumgartner, E. Tunstel, C. Leger, A. Trebi-Ollennu and P. S. Schenker, "Rover Autonomy for Long Range Navigation and Science Data Acquisition on Planetary Surfaces", *Proceedings of the International Conference on Robotics and Automation*, Washington, USA, May 2002.

[11] D. L. Akin, "Science Planning for the Ranger Telerobotic Shuttle Experiment", *Proceedings of the International Symposium on Artificial Intelligence, Robotics and Automation in Space*, Montreal, Canada, June 2001.

[12] M. A. Diftler and R. O. Ambrose, "ROBONAUT: a Robotic Astronaut Assistant", *Proceedings of the International Symposium on Artificial Intelligence, Robotics and Automation in Space*, Montreal, Canada, June 2001.

[13] P. Staritz, S. Skaff, C. Urmson and W. L. Whittaker, "Skyworker: a Robot for Assembly, Inspection and Maintenance of Large Scale Orbital Facilities", *Proceedings of the International Conference on Robotics and Automation*, Seoul, Korea, May 2001, vol. 4, pp. 4180-4185.

[14] T. Williams and S. Tanygin, "On-orbit engineering tests of the AERCam Sprint robotic camera vehicle", *Proceedings of the AAS/AIAA Space Flight Mechanics Meeting*, Monterey, USA, February 1998, pp. 1001-1020.

[15] G. Hirzinger, B. Brunner, J. Dietrich and J. Heindl, "Sensor-Based Space Robotics - ROTEX and Its Telerobotic Features", *IEEE Transactions on Robotics and Automation*, October 1993, vol. 9, no. 5, pp. 649-663.

[16] M. Oda, "ETS-VII, Achievements, Troubles and Future", *Proceedings of the International Symposium on Artificial Intelligence, Robotics and Automation in Space*, Montreal, Canada, June 2001.

[17] R. Mukherji, D. A. Rey, M. Stieber and J. Lymer, "Special Purpose Dexterous Manipulator (SPDM) Advanced Control Features and Development Test Results", *Proceedings of the International Symposium on Artificial*

Intelligence, Robotics and Automation in Space, Montreal, Canada, June 2001.

[18] R. Burridge and J. Graham, "Providing robotic assistance during extra-vehicular activity", *Proceedings of SPIE, Intelligent Systems and Advanced Manufacturing / Mobile Robots XVI*, Boston, USA, October 2001, vol. 4573, pp. 22-33.

[19] E. Zereik, A. Sorbara, G. Casalino and F. Didot, "Autonomous Dual-Arm Mobile Manipulator Crew Assistant for Surface Operations: Force/Vision-Guided Grasping", *International Conference on Recent Advances in Space Technologies*, Istanbul, Turkey, June 2009.

[20] E. Zereik, A. Sorbara, G. Casalino and F. Didot, "Force/Vision-Guided Grasping for an Autonomous Dual-Arm Mobile Manipulator Crew Assistant for Space Exploration Missions", *International Conference on Automation Robotics and Control Systems*, Orlando, USA, July 2009.

[21] J. Bares and D. Wettergreen, "Dante II: Technical Description, Results, and Lessons Learned", *International Journal of Robotics Research*, Palo Alto, USA, July 1999, vol.18, no. 7, pp. 621-649.

[22] C. Harris and M. Stephens, "A Combined Corner and Edge Detector", *Proceedings of the 4th Alvey Vision Conference*, Manchester, England, September 1988.

[23] M. Brown and D. G. Lowe, "Invariant Features from Interest Point Groups", *British Machine Vision Conference*, Cardiff, Wales, September 2002.

[24] H. Bay, T. Tuytelaars and L. van Gool, "SURF: Speeded Up Robust Features", *9th European Conference on Computer Vision*, Graz, Austria, May 2006.

[25] M. Trajkovic and M. Hedley, "Fast Corner Detection", *Image and Vision Computing*, February 1998, vol. 16, no. 2, pp. 75-87.

[26] S. M. Smith and J. M. Brady, "SUSAN Ů A New Approach to Low Level Image Processing", *International Journal of Computer Vision*, May 1997, vol. 23, no. 1, pp. 45-78.

[27] Z. Zhang, "A flexible new technique for camera calibration", *IEEE Transactions on Pattern Analysis and Machine Intelligence*, November 2000, vol. 22, no. 11, pp. 1330-1334.

[28] B. Leroy, G. Medioni, E. Johnson and L. Matthies, "Crater detection for autonomous landing on asteroids", *Image and Vision Computing*, September 2001, vol. 19, no. 11, pp. 787-792.

[29] T. Vinogradova, M. Burl and E. Mjolsness, "Training of a Crater Detection Algorithm for Mars Crater Imagery", *Proceeding of the IEEE Aerospace Conference*, March 2002, Big Sky, USA.

[30] M. Maurette and E. Baumgartner, "Autonomous Navigation Ability: FIDO Test Results", *6th ESA Workshop on Advanced Space Technologies for Robotics and Automation*, Noordwijk, The Netherlands, November 2000.

[31] M. Maimone, Y. Cheng, and L. H. Matthies, "Two Years of Visual Odometry on the Mars Exploration Rovers", *Journal of Field Robotics*, March 2007, vol. 24, no. 3, pp. 169-186.

[32] A. E. Johnson, S. B. Goldberg, Y. Cheng and L. H. Matthies, "Robust and Efficient Stereo Feature Tracking for Visual Odometry", *IEEE International Conference on Robotics and Automation*, Pasadena, USA, May 2008.

[33] J. P. Lewis, "Fast Template Matching", *Conference on Vision Interface*, Quebec City, Canada, May 1995.

[34] G. Casalino and A. Turetta, "Coordination and Control of Multiarm Nonholonomic Mobile Manipulators", MISTRAL: Methodologies and Integration of Subsystems and Technologies for Robotic Architectures and Locomotion, B. Siciliano, G. Casalino, A. De Luca, C. Melchiorri, *Springer Tracts in Advanced Robotics*, Springer-Verlag, April 2004.

[35] G. Casalino, A. Turetta and A. Sorbara, "Dynamic Programming based Computationally Distributed Kinematic Inversion Technique", *Advanced Space Technologies for Robotics and Automation*, Noordwijk, The Netherlands, November 2006.

[36] G. Casalino, A. Turetta and A. Sorbara, "DP-Based Distributed Kinematic Inversion for Complex Robotic Systems", *7th Portuguese Conference on Automatic Control*, Lisbon, Portugal, September 2006.

[37] Richard Hartley and Andrew Zisserman, *"Multiple View Geometry in Computer Vision"*, Cambridge University Press, March 2004.

[38] E. Trucco and A. Verri, *"Introductory Techniques for 3-D Computer Vision"*, Prentice Hall, 1998.

[39] J. Illingworth and J. Kittler, "The Adaptive Hough Transform", *IEEE Transactions on Pattern Analysis and Machine Intelligence*, 1987, vol. 9, no. 5, pp. 690-698.

[40] D. H. Ballard, "Generalizing the Hough transform to detect arbitrary shapes", *Pattern Recognition*, 1981, vol. 13, no. 2, pp. 111-122.

[41] R. D. Braun and R. M. Manning, "Mars Exploration Entry, Descent and Landing Challenges", *Journal of Spacecraft and Rockets*, MarchŰApril 2007, vol. 44, no. 2, pp. 310-323.

[42] I. J. Cox, S. L. Hingorani, S. B. Rao and B. M. Maggs, "A Maximum Likelihood Stereo Algorithm", *Journal of Computer Vision and Image Understanding*, 1996, vol. 63, no. 3, pp. 542-567.

[43] M. Fischler and R. Bolles, "Random Sample Consensus: a Paradigm for Model Fitting with Application to Image Analysis and Automated Cartography", *Communications of the Association for Computing Machinery*, June 1981, vol. 24, pp. 381Ű395.

[44] D. Nistér, O. Naroditsky and J. Bergen, "Visual Odometry", *Proceedings of the IEEE Computer Society Conference on Computer Vision and Pattern Recognition*, Washington, USA, June 2004.

[45] J. A. Robinson, J. L. Rhatigan and D. K. Baumann, "Recent Research Accomplishments on the International Space Station", *Proceedings of the IEEE Aerospace Conference*, Big Sky, USA, March 2005.

[46] J. A. Robinson, T. L. Thumm and D. A. Thomas, "NASA utilization of the International Space Station and the Vision for Space Exploration", *Acta Astronautica*, June 2007, vol. 61, no. 1-6, pp. 176-184.

[47] S. Michaud, M. Dominguez, U. Nguyen, L. Zago and S. Droz, "Eurobot End-Effectors", *Proceedings of the 8th ESA Workshop on Advanced Space Technologies for Robotics and Automation*, Noordwijk, The Netherlands, November 2004.

[48] S. Estable, T. Hülsing, I. Ahrns, H. G. Backhaus, B. Maediger, J. Pizarro De La Iglesia and F. Didot, "Eurobot Monitoring & Control Station", *Proceedings of the International Symposium on Artificial Intelligence, Robotics and Automation in Space*, Hollywood, USA, February 2008.

INDEX